SolidWorks Simulation 2020
Black Book

By
Gaurav Verma
Matt Weber
(CADCAMCAE Works)

Edited by
Kristen

ISBN # 978-1-988722-76-4

NOTICE TO THE READER

DEDICATION

To teachers, who make it possible to disseminate knowledge
to enlighten the young and curious minds
of our future generations

To students, who are the future of the world

THANKS

To my friends and colleagues

To my family for their love and support

Table of Contents

Chapter 3 : Preparing Model for Analysis

Chapter 4 : Static Analysis

Chapter 5 : Non-Linear Static Analysis

Chapter 9 : Thermal Analysis

Chapter 10 : Buckling Analysis

Chapter 11 : Fatigue Analysis

Chapter 16 : Project on Analysis

Practice Questions

Preface

SolidWorks Simulation 2020 is an extension to SolidWorks package. Easy-to-use CAD-embedded analysis capabilities enable all designers and engineers to simulate and analyze design performance. You can quickly and easily employ advanced simulation techniques to optimize performance while you design, to cut down on costly prototypes, eliminate rework and delays, and save you time and development costs.

The **SolidWorks Simulation 2020 Black Book**, is written to help professionals as well as students in performing various tedious jobs of Finite Element Analysis. The book follows a step by step methodology. This book explains the background work running behind your simulation analysis screen. The book covers almost all the information required by a learner to master the SolidWorks Simulation. The book starts with basics of FEA, goes through all the simulation tools and ends up with practical examples of analysis. Chapters on manual FEA ensure the firm understanding of FEA concepts through SolidWorks Simulation. The book contains our special sections named "Why?" and notes. We have given reasons for selecting every option in analysis under the "Why?" sections. The book explains the Solver selection, iteration methods like Newton-Raphson method and integration techniques used by SolidWorks Simulation for functioning. A chapter on Topology Study in this edition helps you understand the procedures of modifying component based on analysis results. New tips and notes have been added in this book for various analyses. Some of the salient features of this book are:

In-Depth explanation of concepts

Every new topic of this book starts with the explanation of the basic concepts. In this way, the user becomes capable of relating the things with real world.

Topics Covered

Every chapter starts with a list of topics being covered in that chapter. In this way, the user can easy find the topic of his/her interest easily.

Instruction through illustration

The instructions to perform any action are provided by maximum number of illustrations so that the user can perform the actions discussed in the book easily and effectively. There are about 700 illustrations that make the learning process effective.

Tutorial point of view

The book explains the concepts through the tutorial to make the understanding of users firm and long lasting. Each chapter of the book has tutorials that are real world projects.

"Why?"

The book explains the reasons for selecting options or setting a parameters in tutorials explained in the book.

Project

Free projects and exercises are provided to students for practicing.

For Faculty

If you are a faculty member, then you can ask for video tutorials on any of the topic, exercise, tutorial, or concept.

Formatting Conventions Used in the Text

All the key terms like name of button, tool, drop-down etc. are kept bold.

Free Resources

Link to the resources used in this book are provided to the users via email. To get the resources, mail us at ***cadcamcaeworks@gmail.com*** or ***info@cadcamcaeworks.com*** with your contact information. With your contact record with us, you will be provided latest updates and informations regarding various technologies. The format to write us e-mail for resources is as follows:

Subject of E-mail as ***Application for resources of _____ Black Book***.
You can give your information below to get updates on the book.
Name:
Course pursuing/Profession:
Contact Address:
E-mail ID:

About Author

The author of this book, Matt Weber, has authored many books on CAD/CAM/CAE books. He has authored **SolidWorks 2020 Black Book** as a CAD companion for this book. **SolidWorks Simulation 2020 Black Book** covers details of simulation and the SolidWorks 2020 Black Book covers all the tools and techniques of modeling. The author has hand on experience on almost all the CAD/CAM/CAE packages. If you have any query/doubt in any CAD/CAM/CAE package, then you can directly contact the author by writing at **cadcamcaeworks@gmail.com**

The technical editor of the book, Gaurav Verma, has authored books on different CAD/CAM/CAE packages. He has authored SolidWorks Flow Simulation 2020 Black Book, SolidWorks Electrical 2020 Black Book, Creo Manufacturing 4.0 Black Book, MasterCAM 2017 for SolidWorks Black Book, AutoCAD Electrical 2020 Black Book, Autodesk Inventor 2018 Black Book, Autodesk Fusion 360 Black Book and many others.

SolidWorks 2020 Black Book can be used as **SolidWorks CAD companion** with this book for learning about modeling tools.

For Any query or suggestion

If you have any query or suggestion please let us know by mailing us on **cadcamcaeworks@gmail.com** or **info@cadcamcaeworks.com**. Your valuable constructive suggestions will be incorporated in our books and your name will be addressed in special thanks area of our books.

This page is left blank intentionally

Chapter 1

Introduction to
Simulation

Topics Covered

The major topics covered in this chapter are:

- *Simulation.*
- *Types of Analyses performed in SolidWorks Simulation.*
- *FEA*
- *User Interface of SolidWorks Simulation.*

SIMULATION

Simulation is the study of effects caused on an object due to real-world loading conditions. Computer Simulation is a type of simulation which uses CAD models to represent real objects and it applies various load conditions on the model to study the real-world effects. SolidWorks Simulation is one of the Computer Simulation programs available in the market. In SolidWorks Simulation, we apply loads on a constrained model under predefined environmental conditions and check the result(visually and/or in the form of tabular data). The types of analyses that can be performed in SolidWorks are given next.

TYPES OF ANALYSES PERFORMED IN SOLIDWORKS SIMULATION

SolidWorks Simulation performs almost all the analyses that are generally performed in Industries. These analyses and their uses are given next.

Static Analysis

This is the most common type of analysis we perform. In this analysis, loads are applied to a body due to which the body deforms and the effects of the loads are transmitted throughout the body. To absorb the effect of loads, the body generates internal forces and reactions at the supports to balance the applied external loads. These internal forces and reactions cause stress and strain in the body. Static analysis refers to the calculation of displacements, strains, and stresses under the effect of external loads, based on some assumptions. The assumptions are as follows.

1. All loads are applied slowly and gradually until they reach their full magnitudes. After reaching their full magnitudes, load will remain constant (i.e. load will not vary against time).
2. Linearity assumption: The relationship between loads and resulting responses is linear. For example, if you double the magnitude of loads, the response of the model (displacements, strains and stresses) will also double. You can make linearity assumption if:

- All materials in the model comply with Hooke's Law that is stress is directly proportional to strain.
- The induced displacements are small enough to ignore the change is stiffness caused by loading.
- Boundary conditions do not vary during the application of loads. Loads must be constant in magnitude, direction and distribution. They should not change while the model is deforming.

If the above assumptions are valid for your analysis, then you can perform **Linear Static Analysis**. For example, a cantilever beam fixed at one end and force applied on other end; refer to Figure-1.

If the above assumptions are not valid, then you need to perform the **Non-Linear Static analysis**. For example, an object attached with a spring being applied under forces; refer to Figure-2.

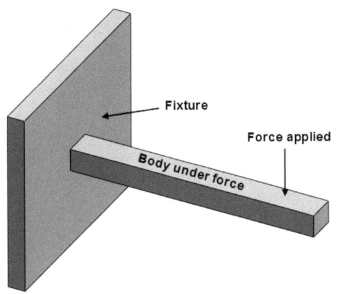

Figure-1. Linear static analysis example

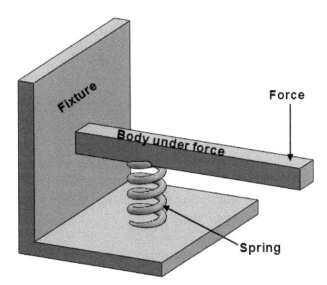

Figure-2. Non-linear static analysis example

Dynamic Analysis

In general, we have to perform dynamic analysis on a structure when the load applied to it varies with time. The most common case of dynamic analysis is the evaluation of responses of a building due to earthquake acceleration at its base. Every structure has a tendency to vibrate at certain frequencies, called natural frequencies. Each natural frequency is associated with a certain shape, called mode shape that the model tends to assume when vibrating at that frequency. When a structure is excited by a dynamic load that coincides with one of its natural frequencies, the structure undergoes large displacements. This phenomenon is known as 'resonance'. Damping prevents the response of the structures to resonant loads. In reality, a continuous model has an infinite number of natural frequencies. However, a finite element model has a finite number of natural frequencies that is equal to the number of degrees of freedom considered in the model. The first few modes of a model (those with the lowest natural frequencies), are normally important. The natural frequencies and corresponding mode shapes depend on the geometry of the structure, its material properties, as well as its support conditions and static loads. The computation of natural frequencies and mode shapes is known as modal analysis. When building the geometry of a model, you usually create it based on the original (undeformed) shape of the model. Some loading, like a structure's self-weight, is always present and can cause considerable changes in the structure's original geometry. These geometric changes may have, in some cases, significant impact on the structure's modal properties. In many cases, this effect can be ignored because the induced deflections are small.

The following few topics – Random Vibration, Response Spectrum analysis, Time History analysis, Transient vibration analysis, and Vibration modal analysis are extensions of dynamic analysis.

Random Vibration

Engineers use this type of analysis to find out how a device or structure responds to steady shaking of the kind you would feel riding in a truck, rail car, rocket (when the motor is on), and so on. Also, things that are riding in the vehicle, such as on-board electronics or cargo of any kind, may need Random Vibration Analysis. The vibration generated in vehicles from the motors, road conditions, etc. is a combination of a great many frequencies from a variety of sources and has a certain "random" nature. Random Vibration Analysis is used by mechanical engineers who design various kinds of transportation equipment.

Response Spectrum Analysis

Engineers use this type of analysis to find out how a device or structure responds to sudden forces or shocks. It is assumed that these shocks or forces occur at boundary points, which are normally fixed. An example would be a building, dam or nuclear reactor when an earthquake strikes. During an earthquake, violent shaking occurs. This shaking transmits into the structure or device at the points where they are attached to the ground (boundary points).

Mechanical engineers who design components for nuclear power plants must use response spectrum analysis as well. Such components might include nuclear reactor parts, pumps, valves, piping, condensers, etc. When an engineer uses response spectrum analysis, he is looking for the maximum stresses or acceleration, velocity and displacements that occur after the shock. These in turn lead to maximum stresses.

Time History Analysis

This analysis plots response (displacements, velocities, accelerations, internal forces etc.) of the structure against time due to dynamic excitation applied on the structure.

Transient Vibration Analysis

When you strike a guitar string or a tuning fork, it goes from a state of inactivity into a vibration to make a musical tone. This tone seems loudest at first, then gradually dies out. Conditions are changing from the first moment the note is struck. When an electric motor is started up, it eventually reaches a steady state of operation. But to get there, it starts from zero RPM and passes through an infinite number of speeds until it attains the operating speed. Every time you rev the motor in your car, you are creating transient vibration. When things vibrate, internal stresses are created by the vibration. These stresses can be devastating if resonance occurs between a device producing vibration and a structure responding to. A bridge may vibrate in the wind or when cars and trucks go across it. Very complex vibration patterns can occur. Because things are constantly changing, engineers must know what the frequencies and stresses are at all moments in time. Sometimes transient vibrations are extremely violent and short-lived. Imagine a torpedo striking the side of a ship and exploding, or a car slamming into a concrete abutment or dropping a coffeepot on a hard floor. Such vibrations are called "shock, " which is just what you would imagine. In real life, shock is rarely a good thing and almost always unplanned. But shocks occur anyhow. Because of vibration, shock is always more devastating than if the same force were applied gradually.

Vibration Analysis (Modal Analysis)

By its very nature, vibration involves repetitive motion. Each occurrence of a complete motion sequence is called a "cycle." Frequency is defined as so many cycles in a given time period. "Cycles per seconds" or "Hertz". Individual parts have what engineers call "natural" frequencies. For example, a violin string at a certain tension will vibrate only at a set number of frequencies, which is why you can produce specific musical tones. There is a base frequency in which the entire string is going back and forth in a simple bow shape.

Harmonics and overtones occur because individual sections of the string can vibrate independently within the larger vibration. These various shapes are called "modes". The base frequency is said to vibrate in the first mode, and so on up the ladder. Each mode shape will have an associated frequency. Higher mode shapes have higher frequencies. The most disastrous kinds of consequences occur when a power-driven device such as a motor for example, produces a frequency at which an attached structure naturally vibrates. This event is called "resonance." If sufficient power is applied, the attached structure will be destroyed. Note that ancient armies, which normally marched "in step," were taken out of step when crossing bridges. Should the beat of the marching feet align with a natural frequency of the bridge, it could fall down. Engineers must design so that resonance does not occur during regular operation of machines. This is a major purpose of Modal Analysis. Ideally, the first mode has a frequency higher than any potential driving frequency. Frequently, resonance cannot be avoided, especially for short periods of time. For example, when a motor comes up to speed it produces a variety of frequencies. So it may pass through a resonant frequency.

Buckling Analysis

If you press down on an empty soft drink can with your hand, not much will seem to happen. If you put the can on the floor and gradually increase the force by stepping down on it with your foot, at some point it will suddenly squash. This sudden scrunching is known as "buckling."

Models with thin parts tend to buckle under axial loading. Buckling can be defined as the sudden deformation, which occurs when the stored membrane(axial) energy is converted into bending energy with no change in the externally applied loads. Mathematically, when buckling occurs, the total stiffness matrix becomes singular. In the normal use of most products, buckling can be catastrophic if it occurs. The failure is not one because of stress but geometric stability. Once the geometry of the part starts to deform, it can no longer support even a fraction of the force initially applied. The worst part about buckling for engineers is that buckling usually occurs at relatively low stress values for what the material can withstand. So they have to make a separate check to see if a product or part thereof is okay with respect to buckling. Slender structures and structures with slender parts loaded in the axial direction buckle under relatively small axial loads. Such structures may fail in buckling while their stresses are far below critical levels. For such structures, the buckling load becomes a critical design factor. Stocky structures, on the other hand, require large loads to buckle, therefore buckling analysis is usually not required.

Buckling almost always involves compression; refer to Figure-3. In mechanical engineering, designs involving thin parts in flexible structures like airplanes and automobiles are susceptible to buckling. Even though stress can be very low, buckling of local areas can cause the whole structure to collapse by a rapid series of 'propagating buckling'. Buckling analysis calculates the smallest (critical) loading required buckling a model. Buckling loads are associated with buckling modes. Designers are usually interested in the lowest mode because it is associated with the lowest critical load. When buckling is the critical design factor, calculating multiple buckling modes helps in locating the weak areas of the model. This may prevent the occurrence of lower buckling modes by simple modifications.

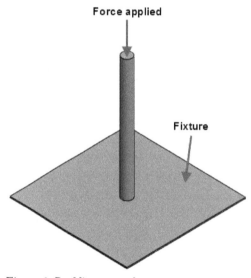

Figure-3. Buckling example

Thermal analysis

There are three mechanisms of heat transfer. These mechanisms are Conduction, Convection, and Radiation. Thermal analysis calculates the temperature distribution in a body due to some or all of these mechanisms. In all three mechanisms, heat flows from a higher-temperature medium to a lower temperature one. Heat transfer by conduction and convection requires the presence of an intervening medium while heat transfer by radiation does not.

There are two modes of heat transfer analysis.

Steady State Thermal Analysis

In this type of analysis, we are only interested in the thermal conditions of the body when it reaches thermal equilibrium, but we are not interested in the time it takes to reach this status. The temperature of each point in the model will remain unchanged until a change occurs in the system. At equilibrium, the thermal energy entering the system is equal to the thermal energy leaving it. Generally, the only material property that is needed for steady state analysis is the thermal conductivity.

Transient Thermal Analysis

In this type of analysis, we are interested in knowing the thermal status of the model at different instances of time. A thermos designer, for example, knows that the temperature of the fluid inside will eventually be equal to the room temperature(steady state), but he is interested in finding out the temperature of the fluid as a function of time. In addition to the thermal conductivity, we also need to specify density, specific heat, initial temperature profile, and the period of time for which solutions are desired.

Till this point, we have learned the basics of various analyses that can be performed in SolidWorks. Now, we will learn about the studies that can be performed in SolidWorks.

Drop Test Studies

Drop test studies simulate the effect of dropping a part or an assembly on a rigid or flexible floor. To perform this study, the floor is considered as planar and flat. The forces that are considered automatically for this study are gravity and impact reaction.

Fatigue Analysis

The fatigue is more over a study then analysis. But it is generally named as analysis. This analysis is used to check the effect of continuous loading and unloading of forces on a body. The base element for performing fatigue analysis are results of static, nonlinear, or time history linear dynamic studies.

Pressure Vessel Design Study

Pressure Vessel Design study allows you to combine the results of static studies with the desired factors and interpret the results. The Pressure Vessel Design study combines the results of the static studies algebraically using a linear combination or the square root of the sum of the squares.

Design Study

Design Study is used to perform an optimization of design. Using the Design Study, you can:

- Define multiple variables using simulation parameters, or driving global variables.
- Define multiple constraints.
- Define multiple goals using sensors.
- Analyze models without simulation results. For example, you can minimize the mass of an assembly with the variables, density and model dimensions, the constraint, and volume.
- Evaluate design choices by defining a parameter that sets bodies to use different materials as a variable.

Till this point, you have become familiar with the analyses that can be performed by using SolidWorks Simulation. But how the software analyze the problems, the answer is FEA.

FEA

FEA, Finite Element Analysis, is a mathematical system used to solve real-world engineering problems by simplifying them. In FEA by SolidWorks, the model is broken into small elements and nodes. Then, distributed forces are applied on each element and node. The cumulative result of forces is calculated and displayed in results. The elements in which a model can be broken into are given in Figure-4,Figure-5,and Figure-6.

Common Finite Elements library for Linear Static and Dynamic Stress Analysis

Element type	Illustration	Description
3-D Truss, 2-nodes		Truss elements are used to provide stiffness between two nodes. These elements transmit compressive and tensile loads along their axis. They do not carry any bending load.
3-D Beam, 2-nodes		Beam elements are used to provide elongational, flexural and rotational stiffness between two nodes. These elements can possess a wide variety of cross-sectional geometries including many standard types
3-D Membrane Plane Stress, 3-nodes		Membrane plane stress elements are used to model "fabric-like" structures, such as tents, cots, domed stadiums, etc. They support three translational degrees of freedom and in-plane (membrane) loading. Orthotropic material properties may be temperature dependent. Incompatible modes are available.
3-D Membrane Plane Stress, 4-nodes		
2-D Elasticity, 3-nodes		Elasticity elements are used for plane strain, plane stress and axisymmetric formulations. They support two translational degrees of freedom. Orthotropic material properties may be temperature dependent. Incompatible modes are available.
2-D Elasticity, 4-nodes		
3-D Brick, 4-nodes		Brick elements are used to simulate the behavior of solids. They support three translational degrees of freedom as well as incompatible displacement modes. Applications include solid objects, such as wheels, turbine blades, flanges, etc.

Figure–4. Finite elements list 1

3-D Brick, 5-nodes		Same as above
3-D Brick, 6-nodes		Same as above
3-D Brick, 8-nodes		Same as above
3-D Plate, 3-nodes		Plate elements are used in the design of pressure vessels, automobile body parts, etc. They support three translational and two rotational degrees of freedom as well as orthotropic material properties. An optional rotational stiffness around the perpendicular axis is automatically added to the node of each element.
3-D Plate, 4-nodes		A thin composite plate element is available for use in models such as mechanical equipment, bicycle frames, etc. A thick composite plate element is also available and can be used in models such as honeycomb sandwich structures, aerospace products, etc. Both thin and thick composite plate elements have no limitations regarding orientation or stacking sequence and support the Tsai-Wu, maximum stress and maximum strain failure criteria.
Tetrahedral, 4-nodes		Tetrahedral elements are used to model solid objects, such as gears, engine blocks and other unusually shaped objects. They support three translational degrees of freedom. They are also available in higher order formulations (mid-side nodes).

Figure-5. Finite elements list 2

Boundary, 2-nodes		Boundary elements are used in conjunction with other elements. A boundary element rigidly or elastically supports a model and enables the extraction of support reactions. Boundary elements are also used to impose a specified rotation or translation.
Gap/Cable, 2-nodes		The gap element simulates compression, where deflection makes two nodes touch and transmit force, such as when a ball bearing moves in a joint. Using gap elements, stresses, bending moments and axial forces where the bearing and joint meet can be determined. A cable element simulates tension, where two nodes moving away from each other a specified distance cause the element to become active. It is still a small-deflection, small-strain analysis, but with deflection-sensitive connectivity.

Figure–6. Finite elements list 3

OPTIMUM PROCESS OF FEA THROUGH SOLIDWORKS SIMULATION

With the knowledge of the basics discussed above we can summarize the process of performing analysis as follows:

1. Construct the part(s) in a solid modeler. It is surprisingly easy to accidentally build flawed models with tiny lines, tiny surfaces or tiny interior voids. The part will look fine, except with extreme zooms, but it may fail to mesh. Most systems have checking routines that can find and repair such problems before you move on to an FEA study. Sometimes, you may have to export a part, and then import it back with a new name because imported parts are usually subjected to more time consuming checks than "native" parts. When multiple parts form an assembly, always mesh and study the individual parts before studying the assembly. Try to plan ahead and introduce split lines into the part to aid in mating assemblies and to locate load regions and restraint (or fixture or support) regions. Today, construction of a part is probably the most reliable stage of any study.

2. Defeature the solid part model for meshing. The solid part may contain features, like a raised logo, that are not necessary to manufacture the part, or required for an accurate analysis study. They can be omitted from the solid used in the analysis study. That is a relatively easy operation supported by most solid modelers (such as the "suppress" option in SW) to help make smaller and faster meshes. However, it has the potential for introducing serious, if not fatal, errors in a following engineering study. This is a reliable modeling process, but its application requires engineering judgment. For example, removing small radius interior fillets can greatly reduces the number of elements and simplifies the mesh generation. But, that creates sharp

reentrant corners that can yield false infinite stresses; refer to Figure-7. Those false high stress regions may cause you to overlook other areas of true high stress levels. Small holes lead to many small elements (and long run times). They also cause stress concentrations that raise the local stress levels by a factor of three or more. The decision to disfeature them depends on where they are located in the part. If they lie in a high stress region you must keep them. But disfeaturing them is allowed if you know they occur in a low stress region. Such decisions are complicated because most parts have multiple possible loading conditions and a low stress region for one load case may become a high stress region for another load case.

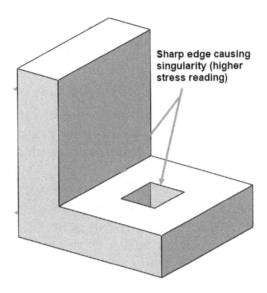

Figure-7. Sharp_edges_causing_singularities.png

3. Combine multiple parts into an assembly. Again, this is well automated and reliable from the geometric point of view and assemblies "look" as expected. However, geometric mating of part interfaces is very different for defining their physical (displacement, or temperature) mating. The physical mating choices are often unclear and the engineer may have to make a range of assumptions, study each, and determine the worst case result. Having to use physical contacts makes the linear problem require iterative solutions that take a long time to run and might fail to converge.

4. Select the element type. Some FEA systems have a huge number of available element types (with underlying theoretical restrictions). The SolidWorks system has only the fundamental types of elements. Namely, truss elements (bars), frame elements (beams), thin shells (or flat plates), thick shells, and solids. The SW simulation system selects the element type (beginning in 2009) based on the shape of the part. The user is allowed to covert a non-solid element region to a solid element region, and visa versa. Knowing which class of element will give a more accurate or faster solution requires training in finite element theory. At times a second element type study is used to help validate a study based on a different element type.

5. Mesh the part(s) or assembly, remembering that the mesh solid may not be the same as the part solid. A general rule in an FEA is that your computer never has enough speed or memory. Sooner or later you will find a study that you cannot execute. Often that means you must utilize a crude mesh(or at least crude in some region) and/or invoke the use of symmetry or anti symmetry conditions. Local solution errors in a

study are proportional to the product of the local element size and the gradient of the secondary variables (i.e., gradient of stress or heat flux). Therefore, you exercise mesh control to place small elements where your engineering judgment estimates high stress (or flux) regions, as well as large elements in low stress regions.

The local solution error also depends on the relative sizes of adjacent elements. You do not want skinny elements adjacent to big ones. Thus, automatic mesh generators have options to gradually vary adjacent element sizes from smallest to biggest.

The solid model sent to the mesh generator frequently should have load or restraint (fixture) regions formed by split lines, even if such splits are not needed for manufacturing the parts. The mesh typically should have refinements at source or load regions and support regions.

A mesh must look like the part, but that is not sufficient for a correct study. A single layer of elements filling a part region is almost never enough. If the region is curved, or subjected to bending, you want at least three layers of quadratic elements, but five is a desirable lower limit. For linear elements you at least double those numbers.

Most engineers do not have access to the source code of their automatic mesh generator. When the mesher fails you frequently do not know why it failed or what to do about it. Often you have to re-try the mesh generation with very large element sizes in hopes of getting some mesh results that can give hints as to why other attempts failed. The meshing of assemblies often fails. Usually the mesher runs out of memory because one or more parts had a very small, often unseen, feature that causes a huge number of tiny elements to be created. You should always attempt to mesh each individual part to spot such problems before you attempt to mesh them as a member of an assembly.

Automatic meshing, with mesh controls, is usually simple and fast today. However, it is only as reliable as the modified part or assembly supplied to it. Distorted elements usually do not develop in automatic mesh generators, due to empirical rules for avoiding them. However, distorted elements locations can usually be plotted. If they are in regions of low gradients you can usually accept them.

You should also note that studies involving natural frequencies are influenced most by the distribution of the mass of the part. Thus, they can still give accurate results with meshes that are much cruder than those that would be acceptable for stress or thermal studies.

6. Assign a linear material to each part. Modern FEA systems have a material library containing the "linear" mechanical, thermal, and/or fluid properties of most standardized materials. They also allow the user to define custom properties. The property values in such tables are often misinterpreted to be more accurate and reliable than they actually are. The reported property values are accepted average values taken from many tests. Rarely are there any data about the distribution of test results, or what standard deviation was associated with the tests. Most tests yield results that follow a "bell shaped" curve distribution, or a similar skewed curve.

When you accept a tabulated property value as a single number to be used in the FEA calculation remember it actually has a probability distribution associated with it. You need to assign a contribution to the total factor of safety to allow for variations from the tabulated property value.

The values of properties found in a material table can appear more or less accurate depending on the units selected. That is an illusion often caused by converting one set of units to another, but not truncating the result to the same number of significant figures available in the actual test units. For example, the elastic modulus of one steel is tabulated from the original test as 210 MPa, but when displayed in other units it shows as 30,457,924.92 psi. Which one do you believe to be the experimental accurate value; the 3 digit value or the 10 digit one? The answer affects how you should view and report stress results. The axial stress in a bar is equal to the elastic modulus times the strain, $\sigma = E\,\varepsilon$. Thus, if E is only known to three or four

significant figures then the reported stress result should have no more significant figures. Material data are usually more reliable than the loading values (considered next), but less accurate that the model or mesh geometries.

7. Select regions of the part(s) to be loaded and assign load levels and load types to each region. In mathematical terminology, load or flux conditions on a boundary region are called Neumann boundary conditions, or nonessential conditions. The geometric regions can be points (in theory), lines, surfaces, or volumes. If they are not existing features of the part, then you should apply split lines to the part to create them before activating the mesh generator. Point forces, or heat sources, are common in undergraduate studies, but in an FEA they cause false infinite stresses, or heat flux. If you include them do not be mislead by the high local values. Refining the mesh does not help since the smallest element still reports near infinite values.

In reality, point loads are better modeled as a total force, or pressure, acting over a small area formed by prior split lines. Saint Venant's Principle states that two different, but statically equivalent, force systems acting on a small portion of the surface of a body produce the same stress distributions at distance large in comparison with the linear dimensions of the portion where the forces act. That also implies that concentrated sources quickly become re-distributed.

In undergraduate statics and dynamics courses engineers are taught to think in terms of point forces and couples. Solid elements do not accept pure couples as loads, but statically equivalent pressures can be applied to solids and yield the correct stresses. Indeed, a couple at a point is almost impossible to create, so the distribution of pressures is probably more like the true situation.

The magnitudes of applied loads are often guesses, or specified by a governing design standard. For example, consider a wind load. A building standard may quote a pressure to be applied for a given wind speed. But, how well do you know the wind speed that might actually be exerted on the structure? Again, there probably is some type of "bell curve" around the expected average speed. You need to assign a contribution to the total factor of safety to allow for variations in the uncertainty of the load value or actual spatial distribution of applied loads.

Loading data are usually less accurate than the material data, but much more accurate that the restraint or supporting conditions considered next.

8. Determine (or more likely assume) how the model interacts with the surroundings not included in your model. These are the restraint (support, or fixture) regions. In mathematical terminology, these are called the essential boundary conditions, or Dirichlet boundary conditions. You cannot afford to model everything interacting with a part. For many decades engineers have developed simplified concepts to approximate surroundings adjacent to a model to simplify hand calculations. They include roller supports, smooth pins, cantilevered (encastre, or fixed) supports, straight cable attachments, etc. Those concepts are often carried over to FEA approaches and can over simplify the true support nature and lead to very large errors in the results.

The choice of restraints (fixations, supports) for a model is surprisingly difficult and is often the least reliable decision made by the engineer. Small changes in the supports can cause large changes in the results. It is wise to try to investigate a number of likely or possible support conditions in different studies. When in doubt, try to include more of the surrounding support material and apply assumed support conditions to those portions at a greater distance from critical part features.

You need to assign a contribution to the total factor of safety to allow for variations in the uncertainty of how or where the actual support conditions occur. That is especially true for buckling studies.

9. Solve the linear system of equations, or the eigenvalue problem. With today's numerical algorithms the solution of the algebraic system or eigen system is usually quite reliable. It is possible to cause ill-conditioned systems (large condition number) with meshes having bad aspect ratios, or large elements adjacent to small ones, but that is unlikely to happen with automatic mesh generators.

10. Check the results. Are the reactions at the supports equal and opposite to the sources you thought that you applied? Are the results consistent with the assumed linear behavior? The engineering definition of a problem with large displacements is one where the maximum displacement is more than half the smallest geometric thickness of the part. The internal definition is a displacement field that significantly changes the volume of an element.

That implies the element geometric shape noticeably changed from the starting shape, and that the shape needs to be updated in a series of much smaller shape changes. Are the displacements big enough to require resolution with large displacement iterations turned on? Have you validated the results with an analytic approximation, or different type of finite element? Engineering judgments are required.

11. Post-process the solution for secondary variables. For structural studies you generally wish to document the deflections, reactions and stresses. For thermal studies you display the temperatures, heat flux vectors and reaction heat flows. With natural frequency models you show (or animate) a few mode shapes. In graphical displays, you can control the number of contours employed, as well as their maximum and minimum ranges. The latter is important if you want to compare two designs on a single page. Limit the number of digits shown on the contour scale to be consistent

with the material modulus (or conductivity, etc.). Color contour plots often do not reproduce well, but graphs do, so learn to use them in you documentation.

12. Determine (or more likely assume) what failure criterion applies to your study. This stage involves assumptions about how a structural material might fail. There are a number of theories. Most are based on stress values or distortional energy levels, but a few depend on strain values. If you know that one has been accepted for your selected material then use that one (as a contribution to the overall factor of safety). Otherwise, you should evaluate more than one theory and see which is the worst case. Also keep in mind that loading or support uncertainties can lead to a range of stress levels, and variations in material properties affect the strength and unexpected failures can occur if those types of distributions happen to intersect as in the next figure.

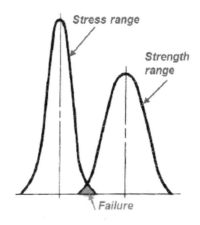

14. Optionally, post-process the secondary variables to measure the theoretical error in the study, and adoptively correct the solution. This converges to an accurate solution to the problem input, but perhaps not to the problem to be solved. Accurate garbage is still garbage.

15. Document, report, and file the study. The part shape, mesh, and results should be reported in image form. Assumptions on which the study was based should be clearly stated, and hopefully confirmed. The documentation should contain an independent validation calculation, or two, from an analytical approximation or an FEA based on a totally different element type.

We will be now use the information discussed above in SolidWorks Simulation. The process of starting SolidWorks Simulation is given next.

STARTING SOLIDWORKS SIMULATION

Before you try to start SolidWorks Simulation, make sure that you have installed it with SolidWorks and you applied the serial key for it during installation. Then, follow the steps given next.

- Start SolidWorks and open the model for which you want to perform the analyses.
- Click on the **SOLIDWORKS Add-Ins** tab and select the **SolidWorks Simulation** button; refer to Figure-8.
- On clicking this button, **Simulation** tab will be added in the **Ribbon**.

- Click on the **Simulation** tab, the **New Study** drop-down will be displayed in the **Ribbon**.

There are two buttons available in this drop-down; **Simulation Advisor** and **New Study**. We will start with **Simulation Advisor** as it is good for novices to the software.

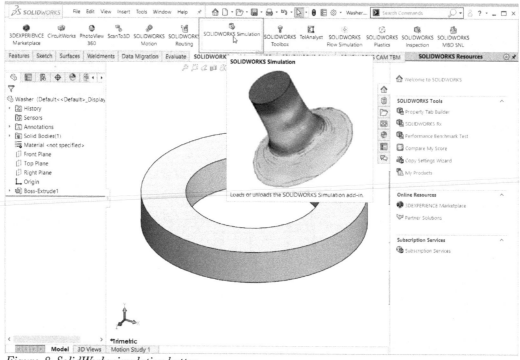

Figure-8. SolidWorks simulation button

On clicking Simulation Advisor

- Click on the **Simulation Advisor** button from the drop-down. The **Simulation Advisor** pane will display in the right of the SolidWorks window; refer to Figure-9. Click on the **Next** button to define type of study. The advisor will be as shown in Figure-10.

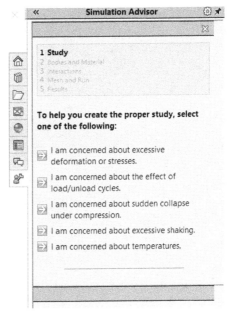

Figure-10. Simulation advisor pane

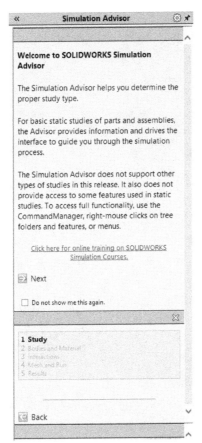

Figure-9. Simulation_Advisor_Task_Pane

- Select the **I am concerned about excessive deformation or stresses** button from the advisor if you want to start a linear or non linear stress study. Select the **I am concerned about the effect of load/unload cycles** button to start a fatigue study. Select the **I am concerned about sudden collapse under compression** button from the advisor if you want to start a buckling analysis. Select the **I am concerned about excessive shaking** button to start frequency study. Select the **I am concerned about temperatures** button from the advisor to start thermal study.
- You will learn more about these options of advisor later in the book.

On clicking New Study

- Click on the **New Study** button from the drop-down. The **Study PropertyManager** will display; refer to Figure-11.

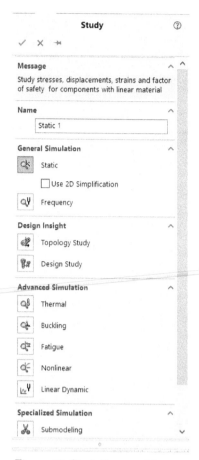

Figure-11. Study PropertyManager

The buttons available in the **Type** rollout of the **PropertyManager** are used to activate different types of analyses.

We will discuss these analyses one by one in the upcoming chapters.

Chapter 2

Basic of Analyses in SolidWork Simulation

Topics Covered

The major topics covered in this chapter are:

- *Starting Analysis*
- *Applying Material*
- *Simulation Evaluator*
- *Defining Fixtures*
- *Applying loads*
- *Defining Connections*
- *Simulating Analysis*
- *Interpreting results*

STARTING ANALYSIS

There are two methods to start an analysis in SolidWorks Simulation; using **Study Advisor** and using **New Study** button. We will use the **New Study** button throughout the book. Although, we will give short notes for using **Study Advisor** to perform the analyses. The steps to start static analysis are given next.

- Click on the **New Study** button from the **Study Advisor** drop-down in the **Simulation** tab of **Ribbon**. The **Study PropertyManager** will display.
- Click on the **Static** button from the **PropertyManager** and specify the name of analysis as desired in the **Name** rollout.
- Click on the **OK** button from the **PropertyManager**. The interface will display as shown in Figure-1.

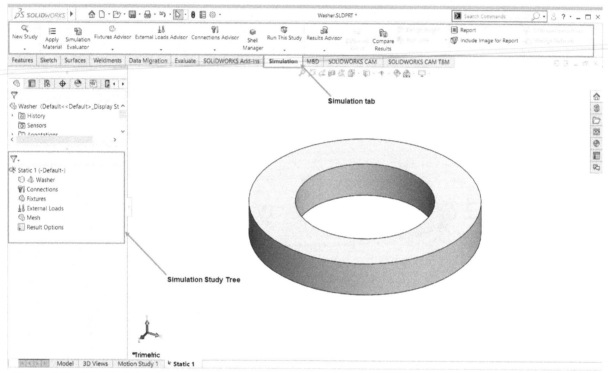

Figure-1. Static analysis interface

The first step for the analysis is to apply material on the model.

APPLYING MATERIAL

Material is a key input for any analysis. The result of analysis is directly related to material of the object. There are few properties of material like, ultimate strength, hardness, and young's modulus which play important role in success/failure of the object under specified load. Also, material determines the application of object in real world. For example, we do not use glass to make pistons in engine. The steps to apply material on object are given next.

- Click on the **Apply Material** button from the **Simulation** tab in the **Ribbon**. The **Material** dialog box will display as shown in Figure-2.

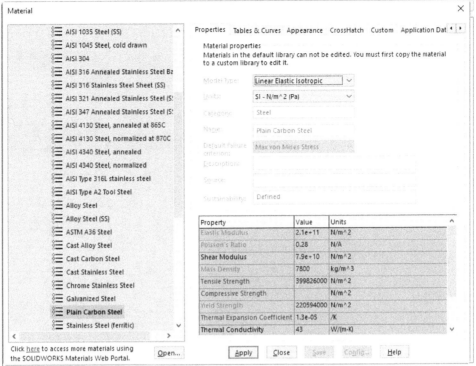

Figure-2. Material dialog box

A library of standard materials is available in this box for selection.

* Select the desired material from left pane of the box. The related parameters will be displayed at the right.
* Click on the **Apply** button to apply the material.
* Click on the **Close** button to close the box.

Although, we have a big library of standard materials available but still there is always a need to custom material. Now, we will discuss about adding custom materials in library.

Adding Custom Materials

Big researches are going on to enhance the properties of engineering materials to make them lighter and stronger. During your job, you may come across a material whose properties are not available in the library. In such a case, follow the steps given next to create custom material.

* Right-click on the **Material** option from the **FeatureManager Design Tree** and select the **Edit Material** option from the shortcut menu displayed; refer to Figure-3. The **Material** dialog box will be displayed as shown in Figure-2 earlier.

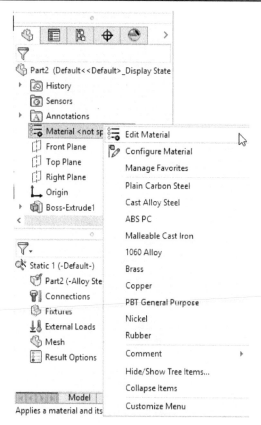

Figure-3. Edit Material option

- Right-click on the material whose properties are close to the material you want to create and select the **Copy** option from the shortcut menu displayed; refer to Figure-4.

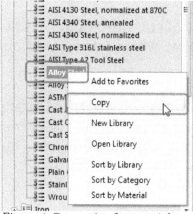

Figure-4. Copy option for material

- Now, move to the bottom of material list and expand the **Custom Material** category; refer to Figure-5.

Figure-5. Custom Material category

- Right-click on the **Custom Materials** category and select the **New Category** option to add a new sub-category in it. Specify the desired name; refer to Figure-6.
- Right-click on the new sub-category and select the **Paste** option from the shortcut menu; refer to Figure-7. The copied material will be added in the sub-category but with the edit-able features. Note that you can start with a completely new material by using the **New Material** option in place of **Paste** option.

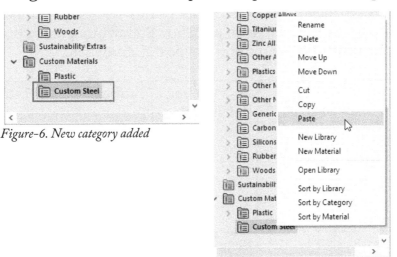

Figure-6. New category added

Figure-7. Paste option for materials

- Click on the added material, the options related to the materials will be displayed on the right in dialog box; refer to Figure-8.
- Click in the required edit boxes and specify the desired values.
- Click on the **Save** button to save the parameters and then click on the **Close** button. Now, you can use this custom material like standard materials.

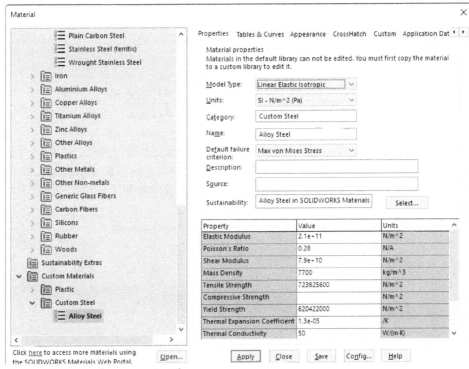

Figure-8. Details of custom material

Role of Material Properties in FEA

While creating new material property, you were asked to specify many different parameters for material like elastic modulus, density, Poisson's ratio, and so on. When we perform FEA on a component, then all the properties are not included in calculations. Depending on the type of analysis, only the related properties are taken in calculations. For example, when you are performing a linear static analysis then thermal expansion coefficient or mass density has no use. You are only concerned about Young's Modulus and Poisson's Ratio. Following table shows the common material properties required for different types of analyses:

	Linear Static	Non-Linear Material	Thermal	Fatigue	Modal & Seismic
Young's Modulus	Required	Required	Required	Required	Required
Poisson's Ratio	Required	Required	Required	Required	Required
Mass Density					Required
Thermal Expansion Coefficient			Required		
Stress-Strain Curve		Required			
Endurance Limit (SN)				Required	

Various properties of material are discussed next.

Young's Modulus or Elastic Modulus or Modulus of Elasticity

The Young's Modulus or Elastic Modulus is the quantity that measures the resistance of material against elastic deformation when stress is applied. Mathematically it can be given as:

$$Young's\ Modulus\ (E) = \frac{Stress}{Strain}$$

Young's Modulus is mainly applicable for tensile elasticity. So, it can be defined as ratio of tensile stress to tensile strain. This parameter is applicable upto yield point of the material in stress strain diagram; refer to Figure-9.

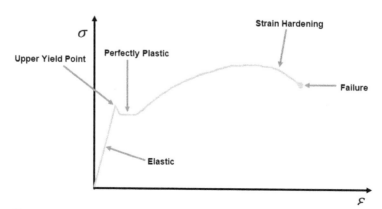

Figure-9. Stress_strain_curve.png

Shear Modulus or Modulus of Rigidity

The Shear Modulus or Modulus of Rigidity (G) is the coefficient of elasticity for shearing force. Shear Modulus is applicable when pure shear is occurring within proportional limit. For linear, homogeneous, and isotropic material; the Modulus of Rigidity can be calculated as :

$$Modulus\ of\ Rigidity\ (G) = \frac{E}{2(1+v)}$$

In simple mathematical term,

Modulus of Rigidity (G) = Shear Stress / Shear Strain

Bulk Modulus

The Bulk Modulus is denoted by K or B. The bulk modulus is used to measure the resistance of an object against compression. This parameter is mainly useful for fluids but it can also be applied on metals when plastic deformation is occurring.

Mathematically, when K > 0

$$Bulk\ Modulus\ (K) = \frac{\Delta P}{\left(\frac{\Delta V}{V}\right)} = -V\frac{dP}{dV}$$

Here, P is pressure and V is volume.

In terms of density, Bulk modulus can be defined as:

$$K = \rho\frac{dP}{d\rho}$$

Here, ρ is density. Note that the inverse of Bulk Modulus gives compressibility of the object.

Density

Density is the mass of an object per unit volume. Density is represented by greek symbol rho (ρ). In terms of equation, it can be represented as:

$$\rho = \frac{m}{V}$$

Poisson's Ratio

The Poisson's ratio defines relation between transverse strain and axial strain of an object under stress. "Poisson's ratio is a measure of the Poisson effect, the phenomenon in which a material tends to expand in directions perpendicular to the direction of compression." - *Wikipedia*. In terms of equation, the Poisson's ratio is define by:

$$\nu = -\frac{d\varepsilon_{trans}}{d\varepsilon_{axial}} = -\frac{d\varepsilon_y}{d\varepsilon_x} = -\frac{d\varepsilon_z}{d\varepsilon_x}$$

Poisson's ratio for steel is 0.27 to 0.30 and for stainless steel is 0.30 to 0.31.

Tensile Strength

Tensile strength is the resistance of material against breaking under tensile force. This parameter is used to define the point at which material goes to plastic deformation (permanent) from elastic deformation (temporary). This value is found experimentally. Tensile strength shows how much tensile stress the material can withstand until it leads to failure in two ways: ductile or brittle failure.

Ductile failure- think of this as the preliminary stage of failure, where it is pushed beyond the yield point to permanent deformation.

Brittle failure- this is the final stage where the tensile strength measurement is taken.

Yield Strength

Yield strength is the maximum stress after which the part starts to deform permanently. This value is also found by experiment.

Compressive Strength

Compressive strength is the resistance of material against permanent size reduction under compression load. Some materials fracture at their compressive strength limit; others deform irreversibly, so a given amount of deformation may be considered as the limit for compressive load. Compressive strength is a key value for design of structures. This value is also found by experiment.

Thermal Expansion Coefficient

The thermal expansion coefficient is the fractional change in size per degree change in temperature. Thermal expansion coefficient can be expressed as linear expansion, area expansion, and volume expansion. The general formula for thermal expansion in case of volume, area, and length is given next.

$$\alpha_V = \frac{1}{V}\left(\frac{\partial V}{\partial T}\right)_p \qquad \frac{\Delta A}{A} = \alpha_A \Delta T \qquad \alpha_L = \frac{1}{L}\frac{dL}{dT}$$

Volume expansion　　　　　Area expansion　　　　　Length expansion

Thermal Conductivity

Thermal conductivity is the measure of a body's capacity to conduct heat energy. It is generally denoted by k. When heat transfer occurs from low temperature body to high temperature body then the amount of heat transfer allowed through the conductor is defined by thermal conductivity. The unit for thermal conductivity is Watt per meter Kelvin- W/(m.K). So, it is the amount of heat transferred by each 1 meter length of conductor for each 1 kelvin temperature difference.

Specific Heat

Specific heat is the amount of heat required to raise the temperature of unit mass substance by one unit. The unit for specific heat is Joule per kg kelvin (J/kg.K).

DEFINING FIXTURES

To perform any analysis, it is important to fix the object by some constrains so that the effect of applied forces can be checked. If you do not apply fixtures then the object will be free to move in the direction of forces applied and no stress or strain will be induced in it. In FEA equation, fixtures are used to reduce the degree of freedoms of elements in the model. There are various options for constraining the object. All these options are explained next.

Fixtures Advisor

This button is used to create fixtures in a guided way. An interactive task pane is used to apply fixtures on the basis of your responses. The steps to use this tool are given next.

*   Click on the **Fixtures Advisor** button from the **Fixtures Advisor** drop-down in the **Ribbon**. The **Simulation Advisor** will display in the right of the window; refer to Figure-10.
*   Click on the **Add a fixture** option from the **Simulation Advisor** pane. The **Fixture PropertyManager** will be displayed; refer to Figure-11.

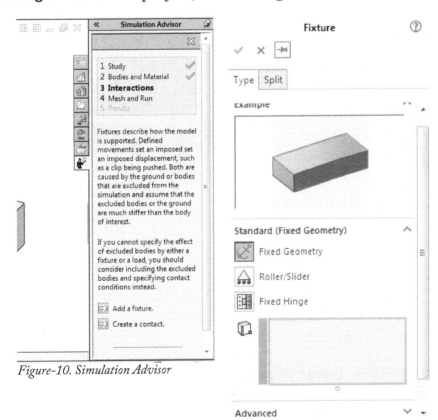

Figure-10. Simulation Advisor

Figure-11. Fixture PropertyManager

Note that the options displayed in the **PropertyManager** can be directly chosen from the **Fixture Advisor** drop-down in the **Ribbon**; refer to Figure-12. Various options in the **PropertyManager** and the drop-down are discussed next.

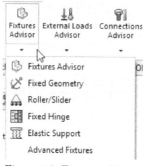

Figure-12. Fixtures Advisor drop-down

Fixed Geometry

This is the most commonly used fixture type. Using this fixture type, you can fix the face, edge or point of the object. To apply this fixture, follow the steps given next.

- Click on the **Fixed Geometry** button from the **Fixtures Advisor** drop-down. The **Fixture PropertyManager** will display as shown in Figure-11.
- Make sure that the **Fixed Geometry** button is selected in the **Standard (Fixed Geometry)** rollout.
- Select the face that you want to be fixed; refer to Figure-13.

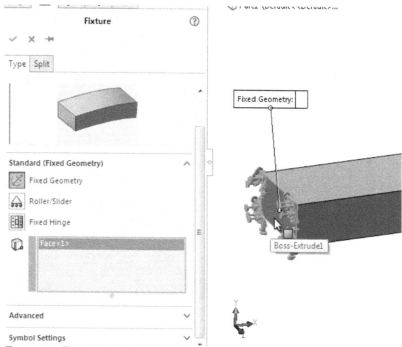

Figure-13. Face selected for fixing

- Click **OK** button to define the fixture.

You can also split a face before clicking **OK** to precisely define the fixed area. Note that splitting faces also changes mesh elements at split line. The steps to do so are given next.

- Click on the **Split** tab from the **Fixture PropertyManager**. The **PropertyManager** will display as shown in Figure-14.
- Click on the **Create Sketch** button from the **PropertyManager**. You are asked to select a plane/face for sketching if not selected earlier for creating fixture.

- Select the desired plane to sketch the splitting line.
- Create the splitting line by using the **Line** tool; refer to Figure-16. Note that you can create any desired shape by using the tools active in the **Ribbon**; refer to Figure-15.

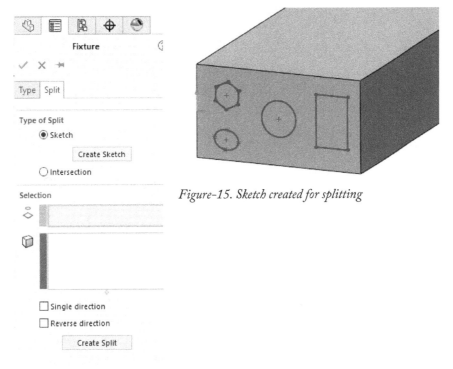

Figure-15. Sketch created for splitting

Figure-14. Fixture Property-Manager with split tab

- Click on the **Exit Sketch** button displayed in **Split** box of viewport. You are asked to select a face.
- Select the faces that you want to split by the projection of drawn sketch line; refer to Figure-17.

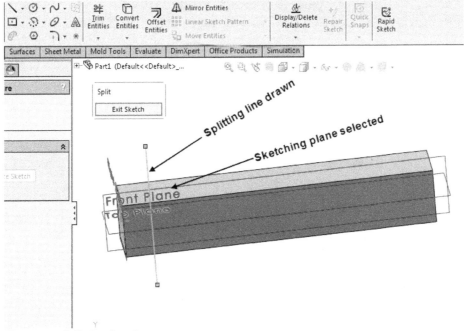

Figure-16. Splitting line drawn

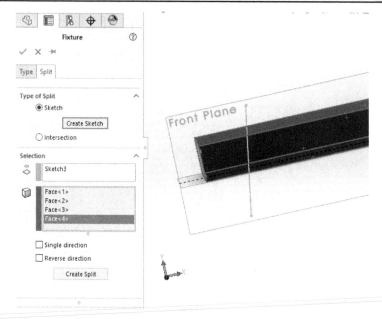

Figure-17. Faces selected for splitting

- Click on the **Create Split** button from the **Selection** rollout of the **PropertyManager**. The selected faces will get split; refer to Figure-18.

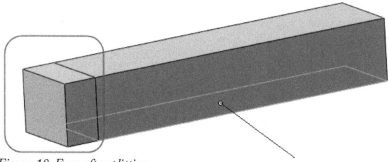

Figure-18. Faces after splitting

- Click on the **Type** tab of the **PropertyManager** to display options related to fixture.
- Click on the faces that you want to fix and remove unwanted faces by right-clicking and then selecting the **Delete** option from the shortcut menu; refer to Figure-19.

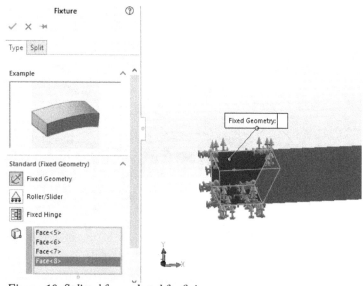

Figure-19. Splitted faces selected for fixing

- Click on **OK** button from the **PropertyManager** to fix the faces.

Where should we use the Fixed Geometry fixture? The answer is: we use this constraint **where we know that the selected face is lying on a hard surface and it cannot move due to action of any applied forces.** A live example can be a building fixed to earth by its base. We expect the building not to move if load increases due to some unexpected guests.

Roller/Slider

This option is used to fix a face in such a way that it can slide/roll in its plane but cannot move perpendicular to the plane. The steps to fix a face by using this option are given next.

- Click on the **Roller/Slider** button from the **Fixtures Advisor** drop-down. The **Fixture PropertyManager** will be displayed similar to the one discussed earlier but with the **Roller/Slider** button selected; refer to Figure-20.
- Select the face that you want to allow for sliding or rolling. Refer to Figure-21.
- Click on the **OK** button from the **PropertyManager** to apply the fixture.

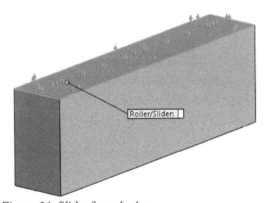

Figure-20. Fixture PropertyManager for roller slider fixture

Figure-21. Slider face selection

 Here the question comes, **where should we use the Roller/Slider fixture?** The answer is: we use this constraint **where object can roll or slide on a surface but can not move normal to it.** A live example can be Tyre of an automobile which rolls on the road and it is expected not to move up towards the sky or down.

Fixed Hinge

This option is used to fix a cylindrical face in such a way that the face act to be hinged. The steps to apply hinge fixture are given next.

- Click on the **Fixed Hinge** tool from the **Fixtures Advisor** drop-down. The **Fixture PropertyManager** will be displayed with the **Fixed Hinge** button selected; refer to Figure-22.
- Select a cylindrical face that you want to fix as hinge; refer to Figure-23.
- Click on the **OK** button to apply the fixture.

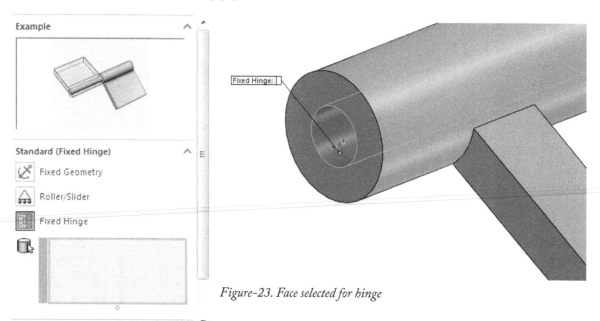

Figure-23. Face selected for hinge

Figure-22. Fixture PropertyManager for fixed hinge fixture

 Where should we use the Fixed Hinge fixture? The answer is: we use this constraint **where object can revolve around its axis.** A live example can be car hood (also called car bonnet in some regions of world).

Elastic Support

This option is used to insert an elastic support between floor and object. This type of fixture is generally applied to rubber or plastic parts. This is more of a connector than fixture. The steps to apply this fixture are given next.

- Click on the **Elastic Support** button from the **Fixtures Advisor** drop-down. The **Connector PropertyManager** will display as shown in Figure-24.

Figure-24. Connectors PropertyManager

- Select the face that you want to make as elastic support; refer to Figure-25.
- Click in the drop-down of **Stiffness** rollout to specify the type of units to be used to specify the value of stiffness.
- Select the desired option from the **SI**, **English (IPS)**, and **Metric (G)** option.
- Click in the **Normal** edit box and specify the stiffness of elastic support in normal direction to the selected face.
- Click in the **Shear** edit box and specify the value of stiffness in the lateral direction (also known as stiffness in shear direction).
- You can specify the total stiffness in normal and shear direction by selecting the **Total** radio button in place of **Distributed** in the **Stiffness** rollout.
- Click on the **OK** button to define the fixture.

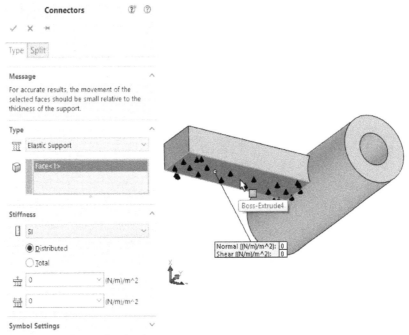

Figure-25. Selecting face for elastic support

Where should we use the Elastic Support fixture? The answer is: we use this constraint **where object is placed or fastened on a soft surface**

like wood or rubber. A live example can be an object placed on wood, rubber or any spring base. I hope you have seen kids jumping on the bed, they rebound because of elastic property of mattress on bed.

Advanced Fixtures

There are some advanced fixtures available in SolidWorks Simulation for constraining the objects. To apply these fixtures follow the steps given next.

- Click on the **Advanced Fixtures** button from the **Fixtures Advisor** drop-down in the **Ribbon**. The **Fixture PropertyManager** will be displayed as shown in Figure-26.

The options available in this **PropertyManager** are discussed next.

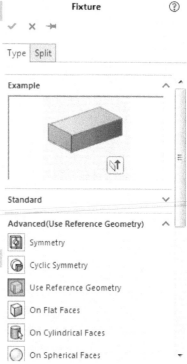

Figure-26. Advanced fixture options

Symmetry

The **Symmetry** button is used to simplify the geometry of the model being analyzed. This option gives us freedom to analyze half of the body and hence reducing size of the problem. Based on this result, the effect is replicated on the other side. This option is not recommended if you are performing buckling or frequency studies. The steps to use this option are given next.

- Click on the **Symmetry** button from the **PropertyManager**. You are asked to select a face to define symmetry.
- Select the face that is symmetric half or quarter of the main part. Refer to Figure-27. Note that you can analyze a part using quarter of it if the part is symmetric and other loading conditions are same for full part. To use quarter part, you will be required to select two faces for defining symmetry; refer to Figure-28.

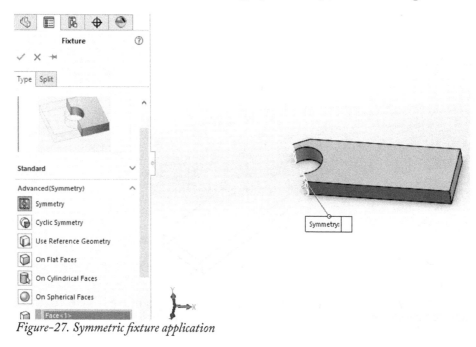

Figure-27. Symmetric fixture application

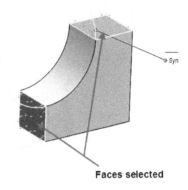

Figure-28. Part symmetry using quarter section

- Click on the **OK** button from the **PropertyManager** to apply the fixture.

 Where should we use the Symmetry fixture? The answer is: we use this constraint **where object is symmetric about a plane or face.** The live example can be a flat table which can be divided into two or four same sections.

Cyclic Symmetry

This button is used to create symmetry for circular objects. Note that the parts that can be used for circular symmetry are generally created by **Revolve** tool and are having an axis for symmetry. The steps to use this option are given next.

- Click on the **Cyclic Symmetry** button from the **PropertyManager**. You are asked to select starting face for circular symmetry; refer to Figure-29. Select the desired face.

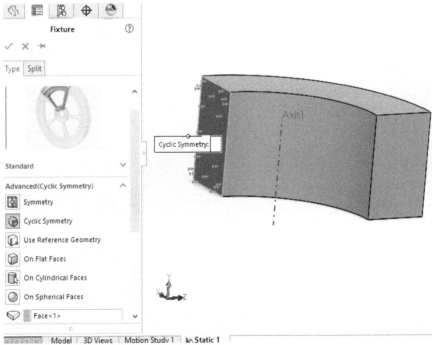

Figure-29. Face selection for cyclic symmetry

- Click in the next selection box in **PropertyManager** and select the ending face for circular symmetry. Similarly, click in the axis selection box in **PropertyManager** and then select the axis about which the part is symmetric. Preview of symmetry will be displayed as shown in Figure-30. Note that the axis should have been created by **Axis** tool in **Reference Geometry** drop-down in the **Analysis Preparation CommandManager**.

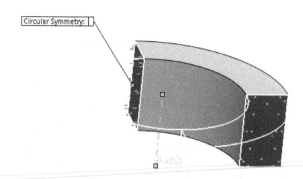

Figure-30. Preview of symmetry fixture

- Click on the **OK** button to apply the circular symmetry fixture.

 Where should we use the Cyclic Symmetry fixture? The answer is: we use this constraint **where object can be created by symmetric blocks sharing the same axis.** A live example can be alloy wheels.

Use Reference Geometry

This button is used to fix the linear/rotational degree of freedom of a body. Depending on the selection of body and reference, you can define whether to fix linear or rotational degree of freedom. To apply this fixture, follow the steps given next.

- Click on the **Use Reference Geometry** button from the **Fixture PropertyManager**. The options in the **PropertyManager** will display as shown in Figure-31.
- Select a face on which you want to apply the fixture. Note that you can also select an edge or vertex for fixture.
- Click in the next selection box in **PropertyManager** and select an edge or face to define the direction reference.
- Click in the **Unit** drop-down in the **Translations** rollout and select the desired unit type.
- Click on the desired direction button from rollout and specify the distance limit to which the selected face can move/rotate.
- The directions for which buttons are not selected, are free to move/rotate.

Figure-31. Fixture PropertyManager with
Use Reference Geometry button selected

On Flat Faces

This button is used to fix a flat face in the desired direction. You can also specify the distance limit in specific directions by using this option. The steps to use this button are given next.

- Click on the **On Flat Faces** button from the **Fixture PropertyManager**. The **PropertyManager** will display as shown in Figure-32.
- Select the face from the model that you want to fix.
- Click on the button of direction in which you want to specify the distance limit.
- Specify the desired distance in the edit box. If you want to fix the body in a direction then specify the value as **0**.
- You can click on more than one button to constrain more directions; refer to Figure-33.

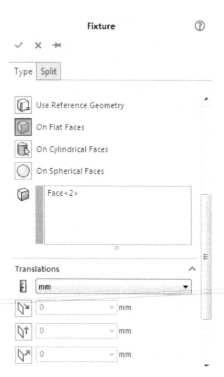

Figure-32. Fixture PropertyManager with On Flat Faces button

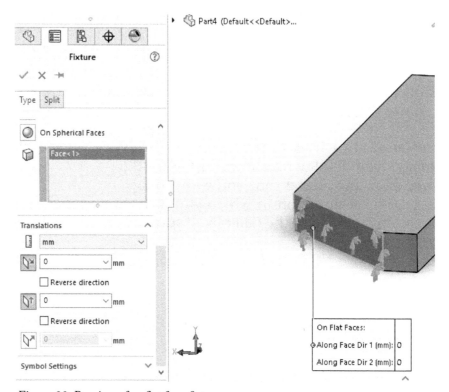

Figure-33. Preview of on flat faces fixture

- Click on the **OK** button from the **PropertyManager** to apply the fixture.

On Cylindrical Faces

This button is used to fix a cylindrical face in the desired direction. This option works in the same way as **On Flat Faces** button works. The steps to use this option are given next.

- Click on the **On Cylindrical Faces** button. The options in the **PropertyManager** will display as shown in Figure-34.
- Select the cylindrical face of the model and select the buttons for the directions for which you want to constrain the body movement.
- Specify the distance/rotation limit for the body for the selected direction buttons. Refer to Figure-35.

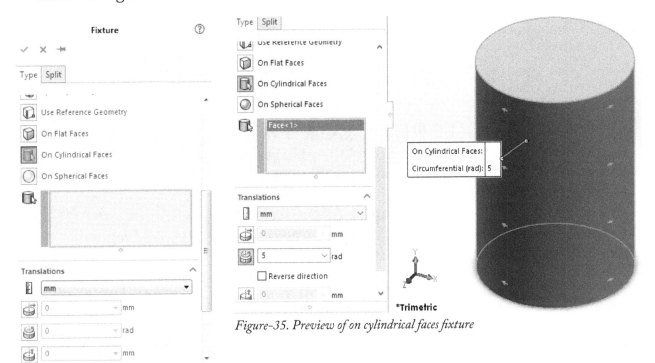

Figure-34. Fixture PropertyManager on selecting On Cylindrical Faces button

Figure-35. Preview of on cylindrical faces fixture

On Spherical Faces

This button is used to fix a spherical face in the desired directions. This option works in the same way as **On Cylindrical Faces** option. The only difference is that you need to select spherical face for **On Spherical Faces** option. Rest of the procedure is same. Refer to Figure-36.

Now, you know all the type of fixtures available in SolidWorks Simulation. The next step is to apply the forces on the objects.

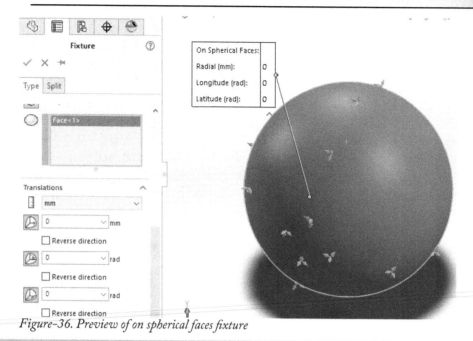

Figure-36. Preview of on spherical faces fixture

EXTERNAL LOAD ADVISOR

The options in this drop-down are used to apply various types of forces on the body. The options in the drop-down are discussed next.

External Loads Advisor

This option is used to display the **Simulation Advisor** with the page related to apply force. The steps to use this option are given next.

- Click on the **External Loads Advisor** option from the **External Loads Advisor** drop-down in the **Simulation** tab of the **Ribbon**. The **Simulation Advisor** will display in the right of the screen as shown in Figure-37.
- Click on the **Add a load** link button from the advisor to apply the load. The options related to forces will displayed in the **Force/Torque PropertyManager**; refer to Figure-38. You can also open the same **PropertyManager** by using the **Force** tool from the **External Loads Advisor** drop-down in the **Ribbon**. The method is discussed next.

Force

This is mostly used option in SolidWorks Simulation for applying loads. To apply the forces, follow the steps given next.

- Click on the **Force** option from the **External Loads Advisor** drop-down in the **Ribbon**. The **Force/Torque PropertyManager** will display as shown in Figure-38.

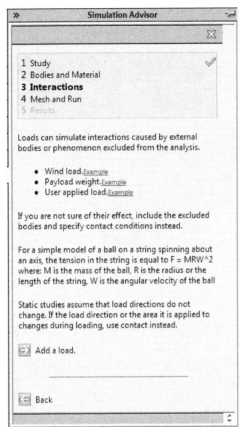

Figure-37. Simulation Advisor with options for load

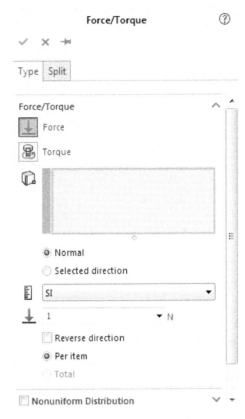

Figure-38. Force Torque PropertyManager

- Select the face on which you want to apply the force.
- To change the direction of force, select the **Selected direction** radio button from the **PropertyManager**. You will be asked to select reference for direction via a pink selection box; refer to Figure-39.
- Click in this selection box and select a reference to specify the direction of force; refer to Figure-40. Note that you need to select the force vector buttons from the **Force** rollout in the **PropertyManager** to apply forces in desired component of selected direction.

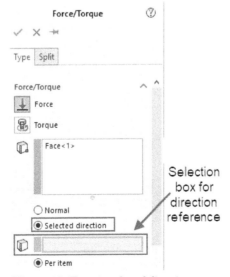

Figure-39. Force in selected direction

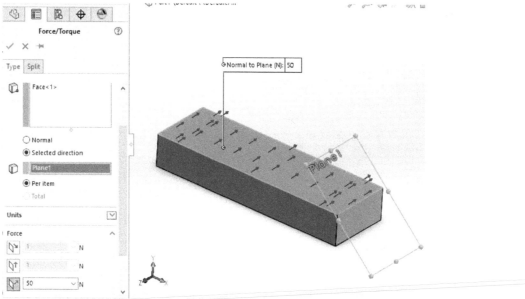

Figure-40. Force applied in selected direction

- If you have selected **Normal** radio button in the **PropertyManager** then click in the **Force Value** edit box and specify the desired value.
- You can change the unit type by selecting desired option from the drop-down in the **Force/Torque** rollout.
- If you have non uniform distribution of force and have an equation of distribution then click on the check box of **Nonuniform Distribution** rollout. The options related to Non-uniform distribution of force will display; refer to Figure-41.

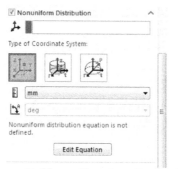

Figure-41. Non uniform force distribution

- There are three coordinate systems to define the equation, Cartesian Coordinate System, Cylindrical Coordinate System, and Spherical Coordinate System. Select the desired button from the **Type of Coordinate System** area of rollout; refer to Figure-42.

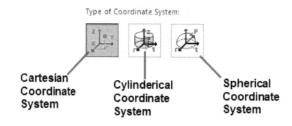

Figure-42. Coordinate system selection

- Click on the **Edit Equation** button from the rollout. The **Edit Equation** dialog box will be displayed as per the selected coordinate system; refer to Figure-43.

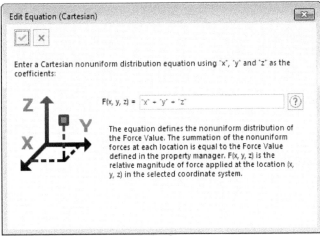

Figure-43. Edit Equation dialog box

Set the desired equation of force in the edit box in **Edit Equation** dialog box. An example of such equation is given next:

$$F(X,Y,Z) = A + B*"X" + C*"Y" + D*"X"*"Y" + E*"X"^\wedge2 + F*"Y"^\wedge2 + G*"Z"^\wedge2$$

- In this equation, you need to enter the numerical values of A, B, C, D, E, F and G.
- Note that the variables of equation which are x, y, and z in case of Cartesian; r, t, and z in case of Cylindrical; and r, t, and p in case of Spherical Coordinate System should be closed in quotation marks ("").
- After specifying the equation, click on the **OK** button from the dialog box.
- Click in the **Coordinate System** selection box and then select the desired coordinate system from the viewport.
- Click on the **OK** button from the **PropertyManager** to apply the force.

Torque

Technically, Torque is product of force and distance. If force is applied on an object in such a way that it tend to rotate the object then it is called torque. The steps to apply torque are given next.

- Click on the **Torque** button from the **External Loads Advisor** drop-down. The **Force/Torque PropertyManager** will be displayed with the **Torque** button selected by default. Refer to Figure-44.
- Select the round face on which you want to apply torque. Preview of the torque will be displayed; refer to Figure-45.

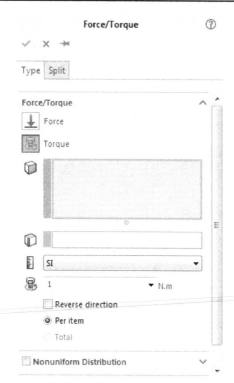

Figure-44. Force Torque PropertyManager with torque options

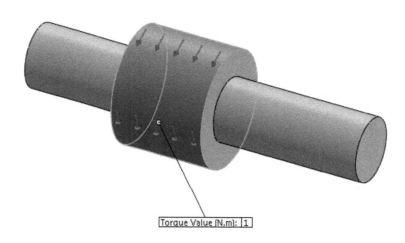

Figure-45. Preview of torque applied

- Specify the desired value of torque in the **Torque Value** edit box of **Force/Torque** rollout.
- The other options in this **PropertyManager** are same as discussed for force. Click on the **OK** button from the **PropertyManager** to apply torque.

Pressure

Pressure is the force applied per unit area. Generally, we apply pressure when we are analyzing pressure vessels or pipes. The steps to apply pressure are given next.

- Click on the **Pressure** option from the **External Loads Advisor** drop-down. The **Pressure PropertyManager** will display as shown in Figure-46.
- Select the desired radio button from the **Type** rollout to specify the direction of pressure. If you select the **Use reference geometry** radio button then one more selection box will be displayed in the **Type** rollout allowing you to select the direction reference.
- Select the face on which you want to apply the pressure and enter the desired value of pressure in the edit box displayed in the **Pressure Value** rollout.
- If you have selected the **Use reference geometry** radio button then select the direction reference to specify the direction of pressure. Preview of the pressure applied will be displayed; refer to Figure-47.
- Click on the **OK** button from the **PropertyManager** to apply the pressure.

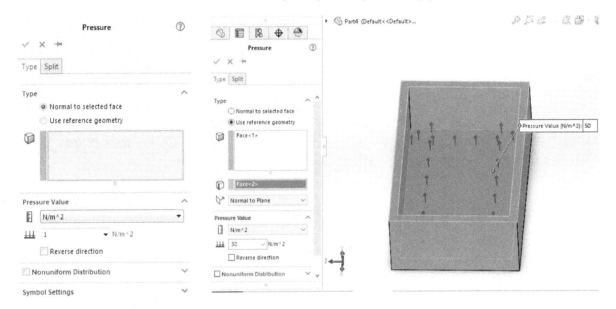

Figure-47. Preview of pressure applied

Figure-46. Pressure PropertyManager

Gravity

Gravity is the force exerted by one object over the other. This is the force that causes objects to fall on the floor on earth. The **Gravity** tool in SolidWorks Simulation is used to apply the gravity force on objects. The steps to use this tool are given next.

- Click on the **Gravity** tool from the **External Loads Advisor** drop-down. The **Gravity PropertyManager** will be displayed as shown in Figure-48.
- Select the plane/face/edge to which define the direction of gravitational force. Preview of the gravitational force will be displayed; refer to Figure-49.
- Select/clear the **Reverse direction** as per your requirement.

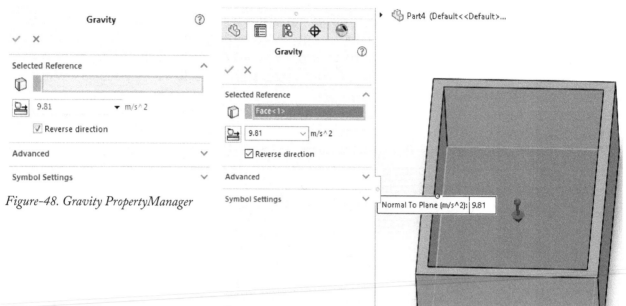

Figure-48. Gravity PropertyManager

Figure-49. Preview of gravitational force

- If you want to specify more components of gravitational force then expand the **Advanced** rollout and enter the values for two other direction vectors of gravitational force; refer to Figure-50.
- Click **OK** to apply the force.

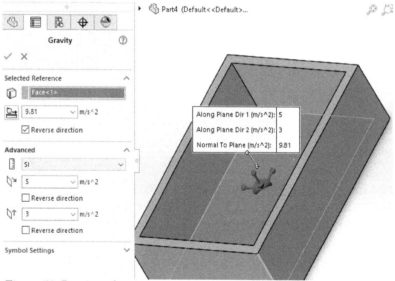

Figure-50. Preview of gravitational force advanced

Centrifugal Force

Centrifugal force is a force which causes the rotating objects to move outward, see a Mary-go-round. If you run a Mary-go-round beyond its design limits then you can see linked seats coming outward. This force is generally generated by rotation of bodies. The steps to apply this force are given next.

- Click on the **Centrifugal Force** button from the **External Loads Advisor** drop-down. The **Centrifugal PropertyManager** will display as shown in Figure-51.

Figure-51. Centrifugal PropertyManager

- Select the round face on which you want to apply the centrifugal force.
- Click in the **Angular Velocity** edit box and specify the angular velocity i.e. omega for the object.
- Click in the **Angular Acceleration** edit box and specify the value of angular acceleration. Preview of the centrifugal force will display; refer to Figure-52.

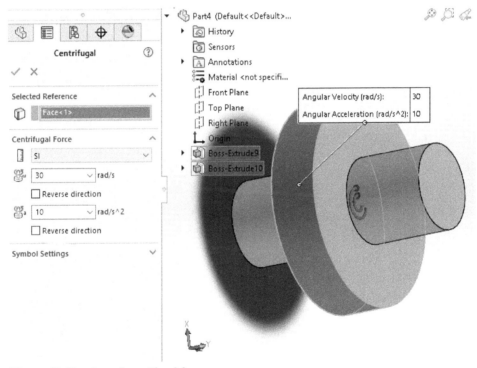

Figure-52. Preview of centrifugal force

- Click on the **OK** button from the **PropertyManager** to apply the force.

Bearing Load

Bearing load is the load exerted by shaft floating in the inner ring of bearing. The bearing load is applied on the round face of bearing and shaft. The steps to apply the load are given next.

- Click on the **Bearing Load** option from the **External Loads Advisor** drop-down. The **Bearing Load PropertyManager** will be displayed as shown in Figure-53.

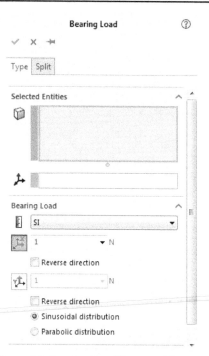

Figure-53. Bearing Load PropertyManager

- Select the faces on which you want to apply the bearing load. Note that you need to select two faces, one from shaft and one from bearing. Also, the selected faces must be half of the full face; refer to Figure-54. Note that you will need the **Split Line** tool to divide the face of bearing and shaft.

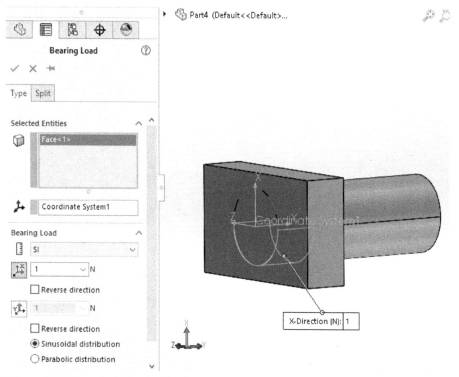

Figure-54. Example for bearing load

- Specify the value of load in X-direction by using the respective edit box in **Bearing Load** rollout.
- If you want to apply load in Y-direction in place of X-direction then click on the **Y-Direction** button and specify the value.

- If you want to change the direction as per your requirement then click in the pink selection box and select the desired coordinate system.
- Select the **Sinusoidal distribution** or **Parabolic distribution** radio button as per your requirement.
- Click on the **OK** button from the **PropertyManager** to apply the bearing load.

Remote Loads/Mass

Remote load/mass is used to apply force on an object in such a way that the origin of force is somewhere else but it is also affecting on the selected face; refer to Figure-55.

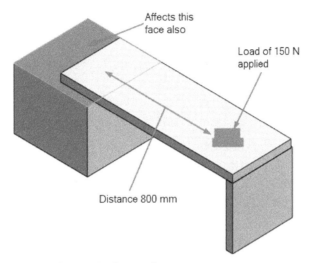

Figure-55. Remote load example

If we apply this load in Solidworks Simulation then it will display as shown in Figure-56.

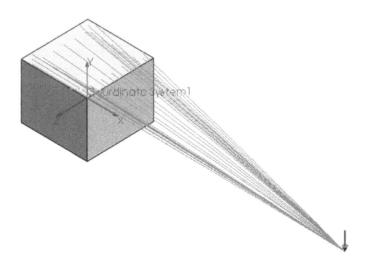

Figure-56. Remote load applied

The procedure to apply remote load is given next.

- Click on the **Remote Load/Mass** option from the **External Loads Advisor** drop-down. The **Remote Loads/Mass PropertyManager** will be displayed as shown in Figure-57.

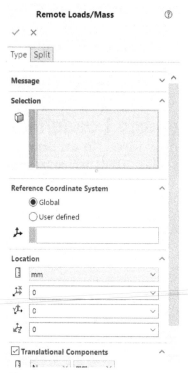

Figure-57. Remote Loads or Mass PropertyManager

- Select the **User Defined** radio button from the **Reference Coordinate System** rollout and then select a reference coordinate system as required. Note that you can select the **Global** radio button to use Global Coordinate System if you do not wish to select any user defined coordinate system. Selected coordinate system will be used as reference for direction.
- Specify the location of remote point by using the edit boxes in the **Location** rollout of the **PropertyManager**.
- Click on the desired button from the **Translational Components** rollout to define the component of force and specify the desired value; refer to Figure-58.
- Select the **Rotational Components** check box to expand the rollout. Specify the desired torque components using the options in this rollout.
- Select the desired radio button from the **Connection Type** rollout to specify the type of remote load/mass to be applied. Note that in case of remote load, the exact force working on the selected face is dependent on the total force applied on the action point and distance of action point from the face. If you select the **Distributed** radio button then the exact amount load which comes after calculation is applied on the selected faces. You can specify the desired weighing factor using the options in the **Weighing Factor** drop-down. If you select the **Rigid** radio button then the load applied on the selected face acts like the load is passed with the help of rigid bars to the selected face.

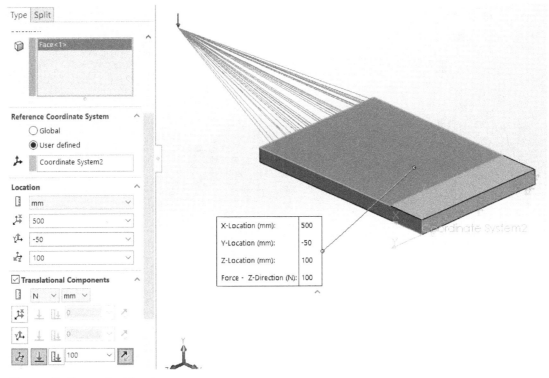

Figure-58. Remote load location

- Select the **Mass** check box to define mass and components of moment of inertia of the object whose load is being transferred.
- Set the other parameters as desired and then click on the **OK** button from the **PropertyManager** to apply the forces.

Distributed Mass

Using this option, you can add the mass of the objects that are not created in the models but their mass has effects on the analysis. The steps to use this option are given next.

- Click on the **Distributed Mass** button from the **External Loads Advisor** drop-down. The **Distributed Mass PropertyManager** will display; refer to Figure-59.

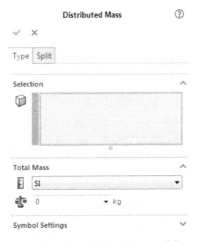

Figure-59. Distributed Mass PropertyManager

- Select the face on which you want to add mass of remote object.
- Click in the **Distributed Mass** edit box and specify the desired value. Preview of the applied mass will be displayed; refer to Figure-60.

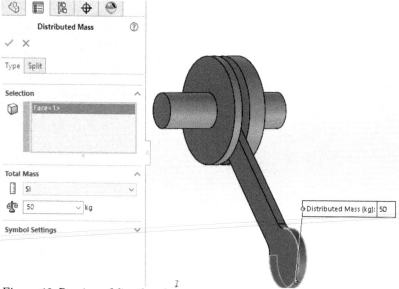

Figure-60. Preview of distributed mass

- Click on the **OK** button from the **PropertyManager** to apply the distributed mass.

Temperature

The **Temperature** option is used to set temperature of the selected face. This option is mostly used in thermal analysis. The steps to use this option are given next.

- Click on the **Temperature** button from the **External Loads Advisor** drop-down. The **Temperature PropertyManager** will display as shown in Figure-61.

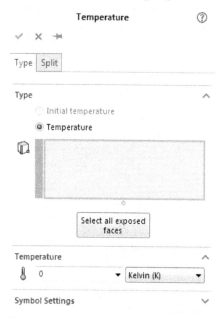

Figure-61. Temperature PropertyManager

- Select the face for which you want to set the temperature.
- Click in the **Temperature** edit box of the **Temperature** rollout and specify the desired value in the edit box. Preview of the applied temperature will be displayed; refer to Figure-62.

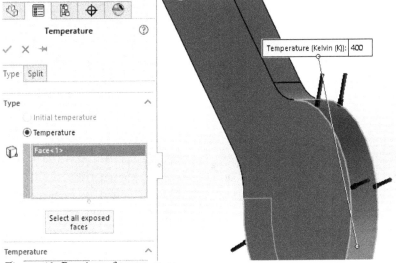

Figure-62. Preview of temperature

- You can apply temperature on all the exposed faces by selecting the **Select all exposed faces** button.

Flow Effects/Thermal Effects

These option are used to add the fluid flow or thermal effects in the current study. The steps to specify the flow effect setting and thermal effect settings are given next.

- Click on the **Flow Effects** or **Thermal Effects** button from the **External Loads Advisor** drop-down. A dialog box will be displayed with the **Flow/Thermal Effects** tab selected; refer to Figure-63.

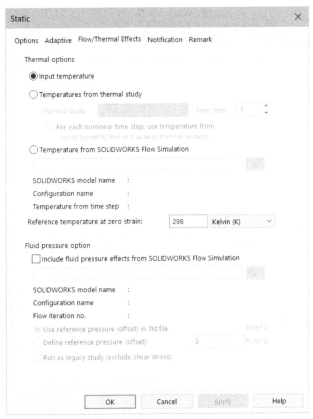

Figure-63. Flow Thermal Effects tab of Static dialog box

- This dialog box is divided into two areas: **Thermal options** and **Fluid pressure options**. The options in these areas are discussed next.

Thermal options

The options in this area are used to specify the source for temperatures in the active study. There are three options to do that:

Input Temperature

Select this radio button to manually specify the temperature for the study. When using this option, make sure to specify temperatures on components or shells. Specifying temperatures on the boundary only may not be practical since a temperature of zero is assumed at all other locations. If you define temperature on boundary only, you may need to create and solve a thermal study first to compute temperatures at all nodes.

Temperatures from thermal study

Select this option to use the output temperature of a thermal study already performed. On selecting this radio button, the options below it will become active. Select the desired thermal study from the drop-down to get is output temperature. You can also set the time for getting the temperature at a specific point of time. For a transient thermal study, select either a Time step number (to import a single temperature) or For each nonlinear time step, use temperature from corresponding time of transient thermal analysis.

Temperatures from SolidWorks Flow Simulation

Select this option to use the output temperature of a completed **SolidWorks Flow Simulation** on the same configuration. Click on this radio button and then click on the **Browse** button to select the result file of flow simulation. Select the desired result file (*.fld) that has been generated by Flow Simulation. Model name, configuration name, and time step number associated with the specified file are displayed.

Reference temperature at zero strain

This edit box is used to set the temperature at which no strains exist in the model.

Fluid pressure option

The options in this area are used to import fluid pressure loads from a SolidWorks Flow Simulation results file.

Include fluid pressure effects from Flow Simulation

Select this option to use the output pressure from a completed SolidWorks Flow Simulation results file. On selecting this option the Browse button will become active below it. Click on the button and select the desired result file (*.fld) that has been generated by Flow Simulation. For proper acquisition of pressure loads from Flow Simulation, the active study and the **Flow Simulation** study should be associated with the same configuration. To create the *.fld file, click **Tools** > **Flow Simulation** > **Tools** > **Export Results to Simulation** tool from the Menu bar once you have activated the **SOLIDWORKS Flow Simulation** add-in and performed flow analysis.

Use reference pressure (offset) in .fld file

Select this option to use the reference pressure defined in the Flow Simulation results file.

Define reference pressure (offset)

Select this option to specify a reference pressure directly. This pressure value is subtracted from the imported pressure values.

Run as legacy study (exclude shear stress)

Using this option, you can import only the normal component of the pressure load from the Flow Simulation.

The **Prescribed Displacement** tool works in the same way as **Use Reference Geometry** tool in **Advanced Fixtures** section discussed earlier. If you are analyzing an assembly, then you need to specify the type of connections between various components of the assembly. The options to specify connections for assembly are available in the **Connections Advisor** drop-down which is discussed next.

CONNECTIONS ADVISOR

The options in this drop-down are used to apply various types of connections between assembly components; refer to Figure-64. You can also specify the connections between components and ground by using these options. These options are discussed next.

Connections Advisor

This option is used to display the page of **Simulation Advisor** that is related to connections between assembly components. The steps to use this option are given next.

- Click on the **Connections Advisor** button from the **Connections Advisor** drop-down in the **Ribbon**. The **Simulation Advisor** will be displayed with the options to apply connections; refer to Figure-65.

Figure-64. Connection Advisor drop-down

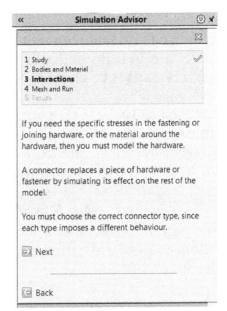

Figure-65. Simulation Advisor with options for connections

- Click on the **Next** button from the **Simulation Advisor**. The list of options that can be used for connections are displayed.
- Select the desired option and follow the instructions given by the software. Note that these options are also available in the **Connections Advisor** drop-down in the **Ribbon**. The options are discussed in the next.

Contact Set

This option is used to specify the contact type between two components or a component and ground. The steps to use this option are given next.

- Click on the **Contact Sets** button from the **Connections Advisor** drop-down. The **Contact Sets PropertyManager** will display as shown in Figure-66.

Figure-66. Contact Sets PropertyManager

- Select the **Automatically find contact sets** option from the **Contact** rollout if you want the software to automatically decide the type of contact to be applied.
- Select the **Manually select contact sets** option to specify the connections manually.
- On selecting the **Manually select contact sets** option, two selection boxes will become available.
- Select the desired option from the drop-down in the **Type** rollout. The options in this drop-down and the process of using these options are given next.

No Penetration

This option is used when you want the faces to be in touch by specified gap value but do not want them to penetrate through each other. The steps to use this option are given next.

- Select the **No Penetration** option from the drop-down in the **Type** rollout. The options in the **PropertyManager** will be displayed as shown in Figure-67.

Figure-67. Options in Contact Sets Property Manager on selecting No Penetration option

- Click in the blue selection box and select the face/edge/point of the first part that you want to specify for connection.
- Click in the pink box and select the face/plane of second component or ground to create the connection with.
- Select the **Friction** check box and specify the coefficient of friction for contact.
- Select the **Gap (clearance)** check box and specify the desired gap between the two components or component and ground.
- You can select the **Self-Contact** check box, if you want your object to not intersect with itself under the application of force; refer to Figure-68.

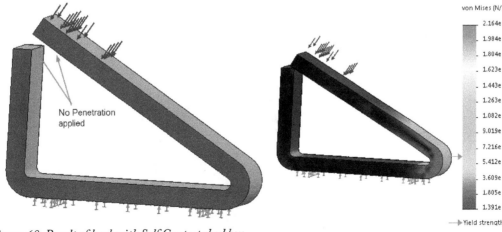

Figure-68. Result of load with Self Contact checkbox

- Click on the check box for **Advanced** rollout and specify the method of contact between the components by selecting the appropriate option.

Bonded

This option is used to attach one component to the other such that both the components are permanently bonded to each other. In other words, effect of force on one component will also be passed on the bonded component perfectly. The steps to use this option are given next.

- Select the **Bonded** option from the drop-down in the **Type** rollout. The options in the **PropertyManager** will be displayed as shown in Figure-69.

Figure-69. Options in Contact Sets Property-Manager on selecting Bonded option

- Click in the blue box and select the face/edge/vertex of the first component.
- Click in the pink box and select the face/plane of the second component. Preview of the bonded connection will be displayed; refer to Figure-70.

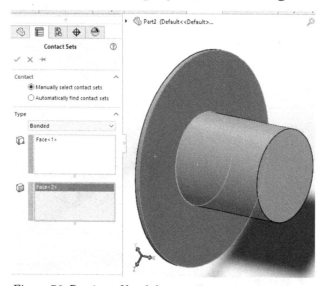

Figure-70. Preview of bonded connection

- Click on the **OK** button from the **PropertyManager** to create the connection.

Sometimes, bonded connection can cause incompatible meshing. This happens because of different size of mesh elements at the connection between two components. Since applying bonded connection means making two components behave like single component so meshing of the two components should be continuous. When the elements on each component are of the same size then you will have compatible mesh. If they are different then you will have an in-compatible mesh. As mentioned earlier this is unique to bonded contacts. With bonded contacts, there exists the opportunity for element nodes at the interface to be merged forming the bonded contact. This is the cleanest way for a bonded contact to be created but isn't always possible. Sometimes the geometry just won't allow for this, for example when the shape and size is dramatically different. In this case, the bonded contact is created through additional contact elements instead of the mesh elements. The solutions for incompatible bonded contact are discussed next.

Using Incompatible Meshing for Assembly

One of the simplest way to deal with incompatible bonded connection is to allow incompatible mesh while meshing. On allowing incompatible mesh, each component is meshed individually with appropriate mesh elements and then all the component meshes are connected with special contact elements. SOLIDWORKS Simulation has two different algorithms to handle the bonds between components with an incompatible mesh – simplified or more accurate. You will learn more about these algorithms later in this book.

Allow Penetration

This option is used to allow penetration of one object into the other object on selected faces. Note that this option is generally used when the objects in assembly are ductile or you are performing frequency and buckling analyses. The steps to used this option are given next.

*   Select the **Allow Penetration** option from the drop-down in the **Type** rollout. The options in the **PropertyManager** will be displayed as shown in Figure-71.

Figure-71. Options in Contact Sets PropertyManager
on selecting Allow Penetration option

- Click in the blue selection box and select the face/edge/vertex of the first component.
- Click in the pink selection box and select the face/plane of the second component.
- Click on the **OK** button from the **PropertyManager**.

Sometimes, you may get an error message about interference between at least two bodies in assembly. You know that there is a need of interference in the assembly and you cannot modify the model. In such cases, the problem is with default contact type which is Bonded. Now, you may ask why it is a problem then answer lies in how meshing will be done for those two bonded bodies. For simple face to face contact of two bonded bodies, the nodes at the contact wall will overlap each other when meshing is created. When there is interference then nodes from first component will overlap wrong nodes of second component which in turn will generate some bad shaped elements very hard to process by FEA. The simple solution to this problem is either use **Allow Penetration** contact or **No Penetration** contact. Note that **No Penetration** is the most time consuming contact type to solve.

Shrink Fit

Shrink fitting is a process in which a shaft is inserted in a hole that is heated up. When the hot hole cools down it strongly arrests the shaft due to contraction. If you need to specify such a connection in the assembly, then this option is used. When there is an interference of more than 0.1% of larger diameter during shrink fit then you need to use this Shrink Fit contact type. For Static studies, you need to use Large displacement option along with this contact type. If there is larger interference then you should use nonlinear study. The steps to use this option are given next.

- Click on the **Shrink Fit** option from the drop-down in the **Type** rollout of **PropertyManager**. The **PropertyManager** will be displayed as shown in Figure-72.

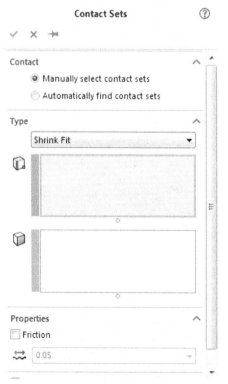

Figure-72. Options in Contact Sets PropertyManager on selecting Shrink Fit option

• You can apply this option in the same way as discussed earlier. You can also specify friction coefficient by selecting the **Friction** check box.

Virtual Wall

Using this option, you can make a component attached to a virtual wall. In this case, the body reacts as if it is in contact with a rigid/flexible wall. This contact type is available for static studies only. Whenever there is a need in static analysis to fix a body against wall using bolted connections the you can use the virtual wall contact. The steps to apply this contact are given next.

• Select the **Virtual Wall** option from the drop-down in the **Type** rollout. The options in the **PropertyManager** will display as shown in Figure-73.

*Figure-73. Options in Contact Sets Property-
Manager on selecting Virtual Wall option*

• Select the face that you want to attach with the virtual wall.
• Click in the violet selection box to collect the plane.
• Select the desired plane that you want to make as virtual wall.
• Click on the **Gap (clearance)** check box to specify the clearance gap.
• Click on the **Rigid** radio button to create a rigid wall and specify the friction coefficient for the wall in the edit box displayed in **Wall Type** rollout.
• Click on the **Flexible** radio button to create a flexible wall.
• Specify the axial stiffness and tangential stiffness in the edit boxes available in the **Wall Stiffness** rollout.
• After setting desired parameters, click on the **OK** button from the **PropertyManager**.

Component Contact

While working on an assembly, this option is generally preferred over the other options available for connections. Using this option, you can specify the contact between two or more bodies along their all shared faces. The steps to use this option are given next.

- Click on the **Component Contact** button from the **Connections Advisor** drop-down. The **PropertyManager** will display as shown in Figure-74.

Figure-74. Component Contact PropertyManager

- Select the desired radio button from the **Contact Type** rollout.
- Select the **Global Contact** check box to apply the selected contact type to all the components of the assembly or select the components one by one to specify the contact type.

No Penetration Contact Type

- If you have selected the **No Penetration** radio button then select the check box of **Friction** rollout and specify the friction coefficient. The friction coefficient specified here, tries to restrict the movement of one component over the other.

Bonded Contact Type

- If you have selected the **Bonded** radio button from the **Contact Type** rollout then you can specify the type of mesh for the components. Select the **Compatible mesh** radio button to create a combined mesh for the all the components bonded. It means there will be a continuous flow of mesh elements at the boundaries of two components. Nodes of elements of first component completely overlap the nodes of elements of second component at the shared boundary; refer to Figure-75. It is not always possible to achieve compatible mesh when there is wide difference between geometries of two components. If you select the **Incompatible mesh** radio button then individual mesh will be created for each of the bonded components and then combined by special elements at boundary; refer to Figure-76. There will be no direct node to node contact between two components. During FEA, the bonded contact is achieved by constraint equation rather then mesh boundary.

Node Overlap between two components

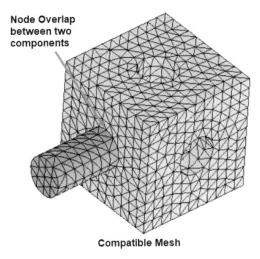

Compatible Mesh

Figure-75. Compatible_mesh.png

No node to node link between components

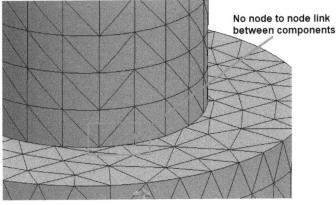

Figure-76. Incompatible_mesh.png

- Select the **Non-touching faces** check box if there is a gap between two components to be bonded. On selecting this check box, the **Maximum Clearance** edit box will be displayed with related unit drop-down. Specify the desired distance value upto which you want two non-touching faces to be assumed as bonded.

There are situations when in an assembly two components are not touching at the moment and have little gap between them but once the simulation starts running, you want these two components to be bonded. In such situations, you need to apply bonded contact type.

- Select the **Include shell edge - solid face/shell face and edge pairs** check box to include shell objects also for creating bonded contact type.
- Click on the **OK** button from the **PropertyManager** to apply the contact.

Allow Penetration Type

Select the Allow Penetration radio button if you want to allow one component to penetrate into another component. This option treats two parts as disjointed. The loads are allowed to cause interference between parts. Using this option can save solution time if the applied loads do not cause interference, so you do not need to apply a No Penetration contact type between parts. Do not use this option unless you are sure that loads will not cause interference on undesired locations.

Contact Visualization Plot

After applying various contacts on components of the assembly, you can review these contacts by selecting **Contact Visualization Plot** button from the **Connections Advisor** drop-down in the **Ribbon**. The steps to use this option are given next.

- Click on the **Contact Visualization Plot** button from the **Connections Advisor** drop-down. The **Contact Visualization Plot PropertyManager** will display as shown in Figure-77.
- Click on the **Calculate** button from the **PropertyManager** to display the connections applied in the assembly. The list of connections will be displayed in the **Results** rollout with preview of contact; refer to Figure-78.

- From SolidWorks 2016 onwards, you can now check the unconstrained bodies in the assembly by using this **PropertyManager**. To do so, click on the **Underconstrained Bodies** tab from the **PropertyManager**. The options will be displayed as shown in Figure-79.

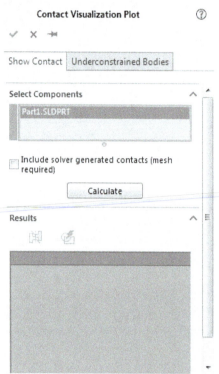

Figure-77. Contact Visualization Plot PropertyManager

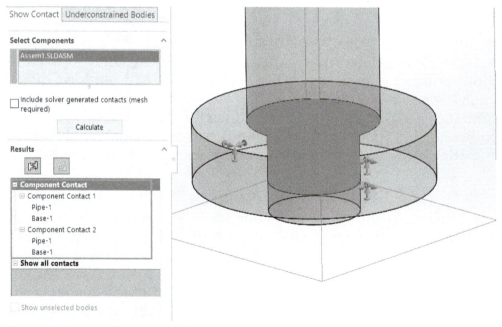

Figure-78. Contact_Visualization_Plot.png

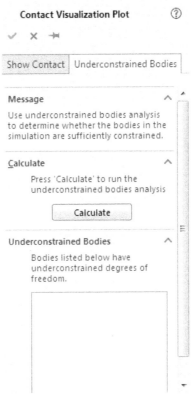

*Figure-79. Contact Visualization Plot PropertyManager
with Underconstrained Bodies tab*

- Click on the **Calculate** button to start analyzing the under-constrained bodies. Once the analysis is done, click one by one on the degree of freedom available for the body in the results area of **PropertyManager** to check the constraining of objects; refer to Figure-80 (In this case, we have not fixed the bottom ring so it is free all 6 degrees of freedom). Note that you must have material properties specified to all the components before using this option.

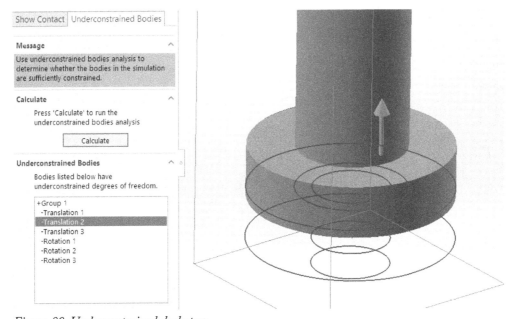

Figure-80. Underconstrained_body.png

- Click on the **OK** button from the **PropertyManager** to exit the tool.

Note that for Global and Component Contact, you do not select specific entities since they apply only to initially touching areas. Use Global Contact to define the most common desired condition and then override it by specifying Component Contacts and Contact Sets wherever needed. So, Global Contact is overruled by Component Contact, and Contact Set will overrule them both. Figure-81 shows the hierarchy of contact definitions.

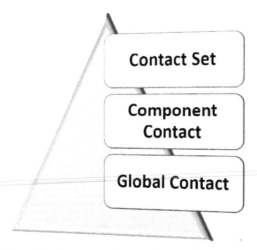

Figure-81. Contact_application_hierarchy.png

Spring

This option is used to connect two components in such a way that they behave like joined by a spring or thick/thin film of lubricant with specified stiffness. The steps to connect two components by using the **Spring** button are given next.

- Click on the **Spring** button from the **Connections Advisor** drop-down. The **Spring PropertyManager** will display as shown in Figure-82.

Figure-82. Connectors Property Manager with Spring option

- Select the desired spring type from the **Type** rollout. There are three type of springs available in the **Type** rollout; **Compression-Extension**, **Compression Only**, and **Extension Only**. When you want to apply spring stiffness against both compression and tensile loads then select the **Compression-Extension** button otherwise select the **Compression Only** or **Extension Only** button to activate respective stiffness.
- Select the desired radio button from **Flat parallel faces**, **Concentric cylindrical faces**, and **Two locations** radio buttons. If **Flat parallel faces** radio button is selected then you need to select two flat parallel faces of the two components between which spring contact is to be applied. Note that most of time, you will need a small section of faces where spring is actually placed in physical model. Select the **Concentric cylindrical faces** radio button to define spring stiffness between two concentric cylindrical faces of two assembly components. Select the Two locations radio button when spring contact is concentrated to selected points only. Note that using two locations radio button for spring contact will generated concentrated stress at selected location in result. You need to interpret the results accordingly.
- Select the entities as per the selected radio button.
- Select the desired unit for stiffness from the **Unit** drop-down in **Options** rollout.
- Select the **Distributed** radio button if you want to define stiffness value per unit area of selected face. Select the **Total** radio button if you want to define total value of stiffness for full selected face. Specify the desired values for normal (compression) stiffness and tangential stiffness in respective edit boxes of the rollout.
- If the spring is under preload then select the **Compression preload force** or **Tension preload force** and specify desired preload in respective edit box.
- Set the other parameters related to spring in the **Options** rollout. Preview of the spring will be displayed; refer to Figure-83.

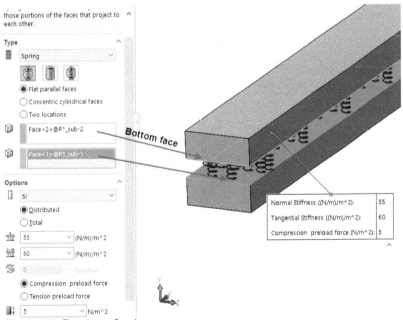

Figure-83. Preview of spring connector

- Click on the **OK** button from the **PropertyManager** to create the spring connection.

Pin

This option is used to connect two cylindrical components with the help of their circular faces. The assemblies that consist of multiple parts can be connected by the pin connector. Examples of assemblies with pins include laptop flap joined with base, scissors lifts, pliers, and actuators. There is also an option to use ring to stop translation of components along the pin or key with pin to stop rotation of components about the pin; refer to Figure-84. The steps to use this option are given next.

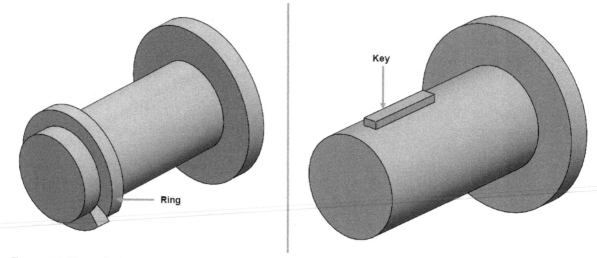

Figure-84. Type_of_pin.png

- Click on the **Pin** button from the **Connections Advisor** drop-down. The **Connectors PropertyManager** will be displayed with the options related to pin connector; refer to Figure-85.

Figure-85. Connectors Property Manager with Pin option

- The selection box is active by default. You need to click on the round faces of the object through which you want to apply the pin; refer to Figure-86. You can select maximum 10 consecutive round faces.

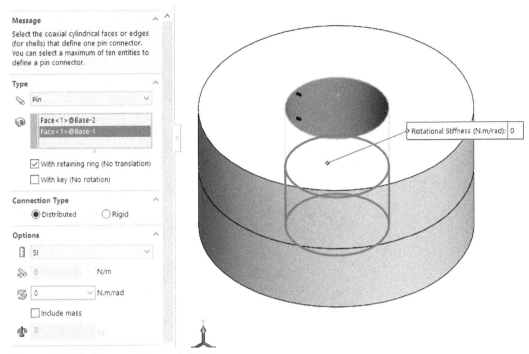

Figure-86. Selecting round faces of object

- Select the desired check boxes from the **Type** rollout to define whether pin is with key, ring, or no accessory.
- Select the **Distributed** radio button from the **Connection Type** rollout if you want the load to be distributed on the pin length. In this way, the stress generated by pin will take the deformed shape of pin into account. Select the Rigid radio button to ignore deformed shape of pin due to load and apply equal stress at connection point due to rigidity. Note that the distributed pin connection is more practical approach if you are concerned about the deformation at connection location.
- Specify the relevant parameters in the **Option** rollout axial stiffness, rotational stiffness, unit system, and so on.
- Select the **Include mass** check box to include the mass of the pin. Specify the mass of pin in the edit box displayed.
- Select the **Strength Data** check box and select the desired material by selecting the **Select material** button.
- Specify the desired tensile stress area of pin to be considered for pin safety. Similarly, specify the pin strength and safety factor in the **Strength Data** rollout. Note that the parameters specified in **Strength Data** rollout are used to check the pin for whether it will pass the test or not.
- After specifying the parameters, click on the **OK** button from the **Connectors PropertyManager** to apply the connection. The pin will be applied and displayed in the model; refer to Figure-87.

Note about Pin Connection: Available for static, nonlinear, buckling, frequency, and dynamic studies. Under loading, pin connectors behave as follows:
- The pin remains straight (it does not bend).
- All faces attached to the pin connector remain coaxial.
- For a rigid connection, each cylindrical face maintains its original shape but can move as a rigid body. For a distributed connection, the coupling nodes of a cylindrical face can move relative to each other.

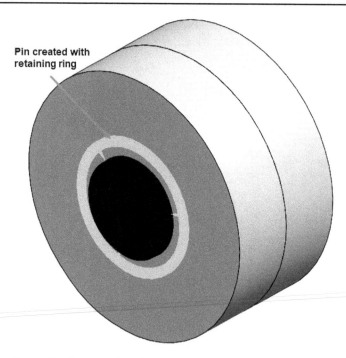

Figure-87. Pin created with retaining ring

Bolt

This option is used to simulate the bolt connection between two components of the assembly. If you have added the bolts in the assembly by using the Toolbox then you can directly convert the bolt into simulation component by using **Yes** option from the dialog box displayed on selecting the **Bolt** option from the **Connections Advisor** drop-down. The steps to use this option are given next.

• Click on the **Bolt** option from the **Connections Advisor** drop-down. If you have added the bolts in the assembly by using the **Toolbox**. Then the **Simulation** dialog box will be displayed as shown in Figure-88. Click on the **Yes** button to automatically apply the simulation connection of bolt.

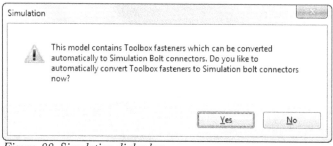

Figure-88. Simulation dialog box

• If the bolt are not taken from the **Toolbox** or you select the **No** button from the **Simulation** dialog box then the **Connectors PropertyManager** will display with the options related bolt connection; refer to Figure-89.

Figure-89. Connectors PropertyManager with Bolt option

- Select the desired button from the **Type** rollout to specify the type of bolt.
- Select the round edge of the bolt head.
- Click in the next selection box in the rollout and select the round edge of the nut. Preview of the bolt connector will be displayed; refer to Figure-90.

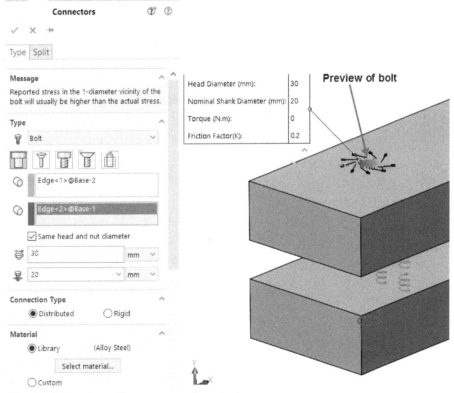

Figure-90. Preview of bot connector after selection

- Set the desired parameters in the **Connection Type** and **Material** rollout as discussed earlier.
- Select the **Strength Data** check box to define strength data as discussed earlier in the **Strength Data** rollout. Select the **Known Tensile Stress Area** radio button

from the rollout to specify the area at which tensile stress is occurring. If you do not know the value then select the **Calculated Tensile Stress Area** radio button and define the thread count. Specify the related parameters using the options given in the rollout displayed.

- You can apply preload on the bolts to simulation the effect of tightened nuts by using the options in the **Pre-load** rollout.
- Expand the **Advanced Options** rollout to define advanced parameters related to bolt connection. Select the **Bolt Series** check box if there are more than two components bounded by same bolt connection and you want to make sure all the components are aligned.
- If the bolt is symmetrical about a plane or face then select the **Symmetrical bolt** check box and select desired symmetry type. Select the **1/2 symmetry** radio button if bolts are half symmetric and select the desired face/plane to define symmetry. Similarly, you can define quarter symmetry by selecting the **1/4 symmetry** radio button.
- Select the **Tight Fit** check box to make the faces in contact with bolt shank rigid.
- Click in the next selection box and select the faces that are in contact with bolt shank.
- After specifying the desired parameters, click on the **OK** button from the **PropertyManager** to apply the connection.

Spot Welds

This option is used to simulate the welding connection between two components at a specific points. Make sure No Penetration contact is applied between faces joined by spot weld. The procedure to use this option are given next.

- Click on the **Spot Welds** option from the **Connections Advisors** drop-down. The **Connectors PropertyManager** will be displayed as shown in Figure-91.
- Select the first face that you want to be welded.
- Click in the next selection box in the **PropertyManager** and select the second face.
- Now, click in the violet selection box and select the point at which both the plates meet each other or where you want the spot weld connector to be applied.

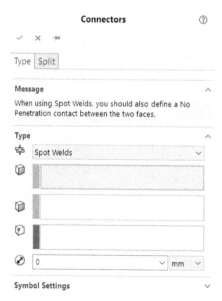

Figure-91. Connectors PropertyManager with Spot Welds option

- Click in the **Spot Weld Diameter** edit box and specify the diameter of the spot weld in the unit selected in drop-down.
- Click on the **OK** button from the **PropertyManager** to create the weld connection.

Edge Weld

This option is used to simulate the welding connection between two components at a common edge. Note that first selection must be a face from shell or sheetmetal part. The procedure to use this option is given next.

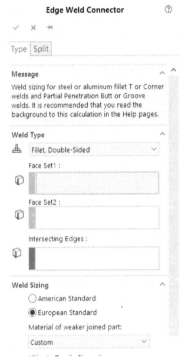

Figure-92. Edge Weld Connector PropertyManager

- Click on the **Edge Weld** button from the **Connections Advisor** drop-down. The **Edge Weld Connector PropertyManager** will be displayed as shown in Figure-92.
- Click on the **Type** drop-down list in the **Weld Type** rollout in **PropertyManager** and select the type of fillet/groove you want to generate on the welding bead.
- Select the first face for welding. Note that you can select only the shell/surface or sheet metal components to apply this connection.
- Click in the next selection box and select the second face for welding.
- Click in the **Intersecting Edges** selection box and select the edge at which the selected two faces intersect.
- You can set the size of welding by using the options in the **Weld Sizing** rollout. You can also switch between American standard and European standard of weld sizing by selecting the corresponding radio button.

Link

This option is used to simulate the connection between two components like the two components are linked at their specified points with a rigid bar. The procedure to use this option is given next.

- Click on the **Link** button from the **Connections Advisor** drop-down. The **Connectors PropertyManager** will be displayed as shown in Figure-93.
- Click on the desired point of first entity.
- Click in the violet colored selection box and select the second point of link on the second entity.
- Click on the **OK** button to create the link. Note that on specifying the link connector, the distance between the two selected points will remain the fixed.

Bearing

This option is used to simulate the connection between two components like the two components are linked to each other with the help of a bearing. This connection is generally applied to shafts and housings. The procedure to use this option is given next.

- Click on the **Bearing** button from the **Connections Advisor** drop-down. The **Connectors PropertyManager** will be displayed with the options related to bearing connection; refer to Figure-94.

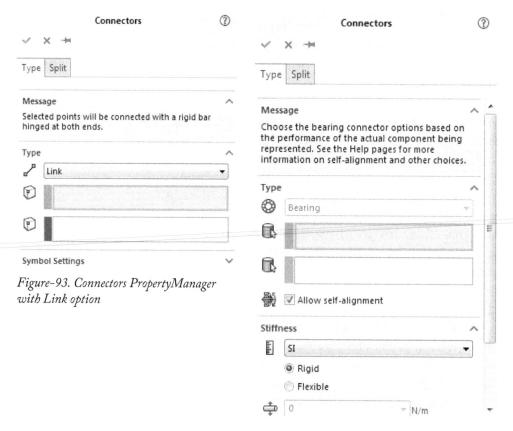

Figure-93. Connectors PropertyManager with Link option

Figure-94. Connectors PropertyManager with bearing option

- Select the round face of the shaft.
- Click in the next selection box in the **Type** rollout and select the round face of the housing.
- Specify the desired parameters in the **Stiffness** rollout and click on the **OK** button from the **PropertyManager** to apply the connection. See Figure-95.

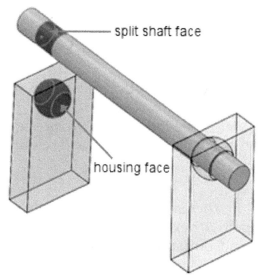

Figure-95. Preview of bearing connection

Rigid Connection

This option is used to apply rigid connection between the two components. The procedure to create the rigid connection is given next.

- Click on the **Rigid Connection** button from the **Connections Advisor** drop-down. The **Connectors PropertyManager** will be displayed as shown in Figure-96.

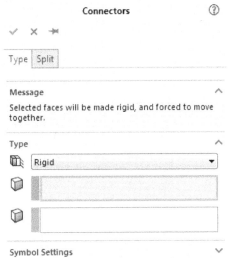

Figure-96. Connectors PropertyManager with rigid option

- Select the first face for applying rigid connection.
- Click in the next selection box and select the second adjoining face.
- Click on the **OK** button from the **PropertyManager** to create the connection.

After specifying the desired connections for the assembly/component, the next step is to create and refine mesh for the assembly/component. The methods to create mesh are discussed next.

MESHING

Meshing is the base of FEM (Finite Element Method) which is one of the methods used for FEA. Meshing divides the solid/shell models into elements of finite size and shape. These elements are joined at some common points called nodes. These nodes define the load transfer from one element to other element. Meshing is a very crucial step in design analysis. The automatic mesher in the software generates a mesh based on a global element size, tolerance, and local mesh control specifications. Mesh control lets you specify different sizes of elements for components, faces, edges, and vertices.

The software estimates a global element size for the model taking into consideration its volume, surface area, and other geometric details. The size of the generated mesh (number of nodes and elements) depends on the geometry and dimensions of the model, element size, mesh tolerance, mesh control, and contact specifications. In the early stages of design analysis where approximate results may suffice, you can specify a larger element size for a faster solution. For a more accurate solution, a smaller element size may be required.

Meshing generates 3D tetrahedral solid elements, 2D triangular shell elements, and 1D beam elements. A mesh consists of one type of elements unless the mixed mesh type is specified. Solid elements are naturally suitable for bulky models. Shell elements are naturally suitable for modeling thin parts (sheet metals), and beams and trusses are suitable for modeling structural members.

The procedure to create the mesh of the solid is given next.

- Click on the **Create Mesh** button from the **Run** drop-down in the **Ribbon**. The **Mesh PropertyManager** will be displayed as shown in Figure-97.
- Move the slider to make the elements of mesh fine or coarse. If you move the slider towards the right, the elements of the mesh will become fine. If you move the slider towards the left, the mesh will become coarse.
- Select the **Mesh Parameters** check box to specify the parameters for mesh.
- Click on the **Curvature based mesh** radio button if you want to create a curve based mesh.
- Click on the **Blended curvature-based mesh** radio button if you want to mesh the object along blended curvature. Figure-98 shows meshing of a ring by three methods.

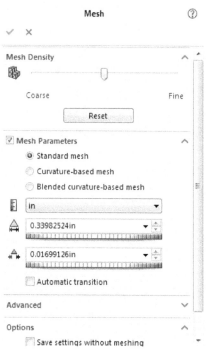

Figure-97. Mesh PropertyManager

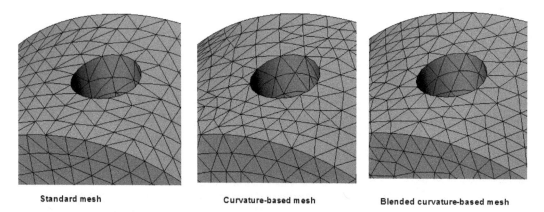

Standard mesh   Curvature-based mesh   Blended curvature-based mesh

Figure-98. Mesh types

- Select the **Automatic transition** to apply mesh controls to small features, holes, fillets, and other fine details of your model. Clear **Automatic transition** before meshing large models with many small features and details to avoid generating large numbers of elements.
- Click on the **Advanced** rollout button to expand display advanced options related to the mesh.
- Select the **Draft** option from the **Quality Mesh** drop-down to set the quality of mesh for fast evaluation of the analysis. If you select this option then 4 corner nodes for solid elements (tetrahedral elements) and 3 corner nodes for shell element (triangular elements) are created. In draft quality mesh, first order mesh elements are used. Select the **High** option from the drop-down to use second order elements.
- The **Jacobian Points** are used to specify the number of integration points for checking distortion while performing the analysis. You can select 4, 16, or 29 Gauss points or you can select the **At Nodes** option.
- The **Automatic trails for solid** check box is selected to find out and apply the optimum mesh size automatically. Note that the ratio by which the global element size and tolerance are reduced for each trial is **0.8**. On selecting this check box, the **Number of trials** spinner is displayed. Using this spinner, you can specify the maximum number of trials for re-meshing.
- Select the **Render shell thickness in 3D (slower)** check box to display shell bodies with their thickness as 3D representation of shell elements.
- If your part fails while meshing at some portions, then select the **Remesh failed parts with incompatible mesh** check box to create incompatible mesh at the portions that failed.
- If you do not want to mesh now but want to save the settings then select the **Save settings without meshing** check box.
- To run the analysis directly after creation of mesh, select the **Run (solve) the analysis** check box.
- Click on the **OK** button from the **PropertyManager** to apply settings.

Figure-99 shows the meshing of a model with the loads and fixtures applied.

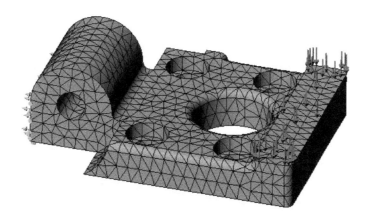

Figure-99. Preview of mesh with loads applied

SIMULATION EVALUATOR

The Simulation Evaluator tool is used to check whether all the parameters have been specified for running current analysis. The procedure to use this tool is given next.

- Click on the **Simulation Evaluator** tool from the **Simulation CommandManager** in the **Ribbon**. The **Simulation Evaluator** dialog box will be displayed; refer to Figure-100.

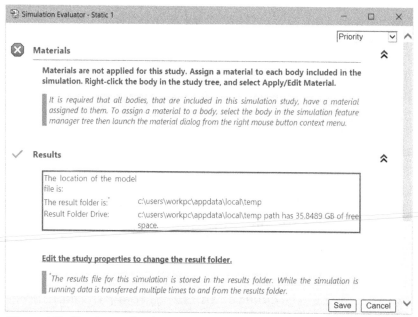

Figure-100. Simulation_Evaluator_dialog_box.png

- This evaluator checks for material, location of results file, and mesh parameters. Note that this evaluator does not check for loads and constraints. You need to check for loads and constraints manually. The parameters that have not been defined will be displayed with a red cross mark and all the parameters that are correct for analysis are with green tick mark.
- Click on the Save button to save the result file at desired location.

RUNNING ANALYSIS

After we have applied all the parameters related to analysis and we have created the mesh, the next step is to run the analysis. The steps to run the analysis are given next.

- Click on the **Run** button from the **Ribbon**. The analyzing process will start and the result will be displayed; refer to Figure-101. There is no setting and nothing you can do during this process but to wait!
- After the analysis is complete, the next step is to interpret the results of the analysis. To interpret the results for various purposes, the tools are available in the **Result Advisors** drop-down which is discussed next.

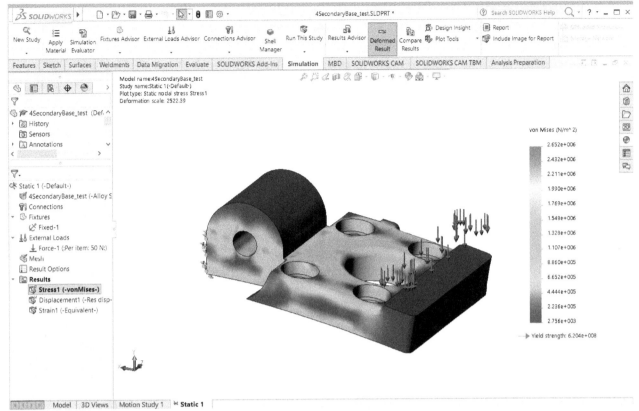

Figure-101. Preview of simulation result

RESULTS ADVISOR

The tools in this drop-down are available only after you successfully running the desired type of analysis. The tools that get available after running static analysis are given next.

- Results Advisor
- Model Only (No Results)
- New Plot menu
- List Stress,Displacement and Strain
- List Result Force
- List Contact Force
- List Pin\Bolt\Bearing Force

The tools above are discussed next.

Results Advisor

This tool is used to display the result page of the **Simulation Advisor**; refer to Figure-102. The steps to use the options in this page of Simulation Advisor are given next.

- Click on the **Play** button from the **Simulation Advisor** to play the simulation of the applied analysis. Note that the limits of deflection are displayed in the **Simulation Advisor**. For example, in Figure-102, the maximum deflection is **2.29361e-005m**.
- After you have checked the simulation, you need to change the parameters of model to get the desired results. For example, you want to change the material

to increase the stiffness of the model. To perform these changes, we are provided with the three link buttons in the **Simulation Advisor**.

*   Click on the desired link button from the first three link buttons to change the parameters. We will discuss these options later in the book.

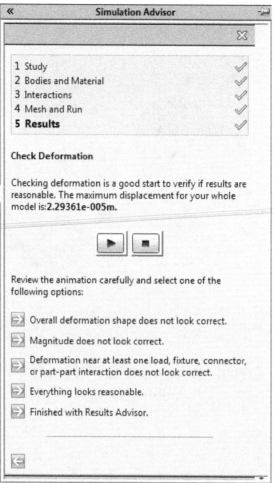

Figure-102. Results page Simulation Advisor

Model Only(No Results)

This tool is used to display the model without displaying any result of analysis. The procedure to use this tool is discussed next.

*   Click on the **Model Only (No Results)** button from the **Results Advisor** drop-down in the **Ribbon**. The model will be displayed.

New Plot

The options in **New Plot** cascading menu are used to generate new plots for various parameters on the basis of performed analysis. In case of a Static analysis, you can get the options as shown in Figure-103.

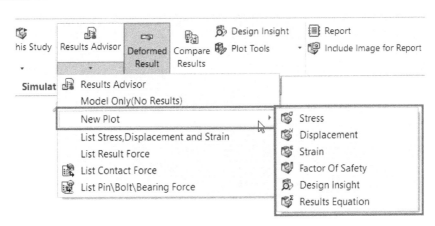

Figure-103. Options in New Plot cascading menu

The Stress, Strain and Displacement plots are added in the results automatically. The procedure to add Factor of Safety will be discussed in next chapter. The procedures to add Design Insight, and Results Equation plots are given next.

Design Insight

The Design Insight plot is used to display the areas that can bear the load most efficiently. In simple words, the area that are under minimum stress are displayed by this tool.

• Click on the **Design Insight** option from **New Plot** cascading menu. The respective **Design Insight PropertyManager** will be displayed; refer to Figure-104.

Figure-104. Design Insight Property-Manager

• Specify the desired settings in **Settings** tab of the **PropertyManager** and click on the **OK** button to display the desired type of analysis result; refer to Figure-105. Note that the link for selected analysis result will be added in the **Results** node in the left area.

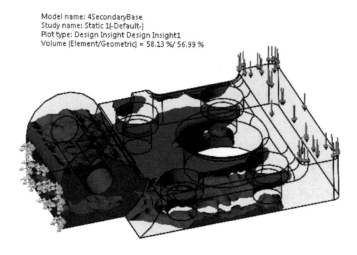

Model name: 4SecondaryBase
Study name: Static 1(-Default-)
Plot type: Design Insight Design Insight1
Volume (Element/Geometric) = 58.13 %/ 56.99 %

Figure-105. Preview for design insight

Results Equation

- Click on the **Results Advisor -> New Plot -> Results Equation** option from the **Ribbon**. The **Results Equation Plot PropertyManager** will be displayed along with the **Edit Equation** dialog box; refer to Figure-106.

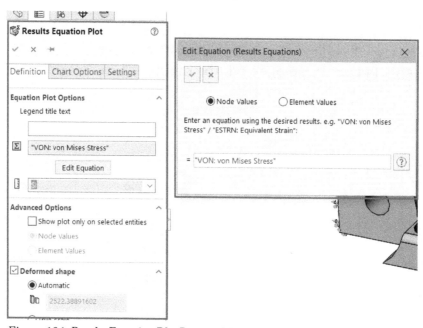

Figure-106. Results Equation Plot PropertyManager with Edit Equation dialog box

- Select the **Node Values** radio button or the **Element Values** radio button from the dialog box to make equations for nodes or elements in the model respectively.
- Specify desired formula in the edit box like, **"EPSX: X Normal Strain"+"EPSY: Y Normal Strain"+"EPSZ: Z Normal Strain"** to check triaxial strain on each node or element in the result. Note that when you type a mathematical operator or click in the empty edit box then a flyout is displayed with equation parameters; refer to Figure-107. You can select these parameters for creating the equation.

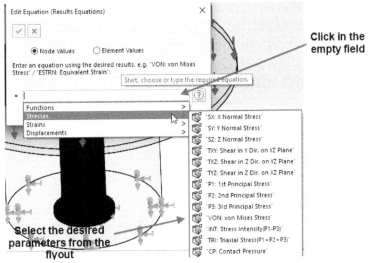

Figure-107. Flyout with equation parameters

- After creating equation, click on the **OK** button from the dialog box.
- Enter the desired name of plot in the **Legend title text** edit box of **Equation Plot Options** rollout of the **PropertyManager**.
- Select the desired unit system from the **Units** drop-down in the rollout.
- Set the other parameters as required and then click on the **OK** button from the **PropertyManager**. The new plot will be created; refer to Figure-108. Note that the new plot will be added in the **Results** node at the right of the dialog box.

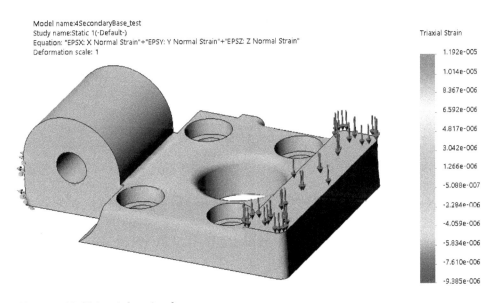

Figure-108. Tri-axial strain plot

List Stress, Displacement and Strain

This tool is used to display the values of stress, displacement, and strain at node intersecting with the selected plane. The procedure to use this option is given next.

- Select the plane by which you want to check values of stress, displacement and strain at the intersecting nodes.
- Click on the **List Stress, Displacement and Strain** tool from the **Results Advisor** drop-down. The **List Results PropertyManager** will be displayed; refer to Figure-109.

- Select the desired radio button from the **Quantity** rollout. The options related to the selected radio button will be displayed in the **Component** rollout. Select the desired component of stress/displacement/strain from the **Component** drop-down in the rollout.

- Click on the **Advanced Options** to expand the rollout. The advanced options related to the selected radio button will be displayed. Refer to Figure-110.

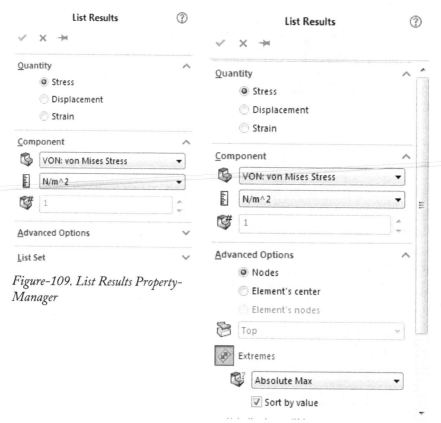

Figure-109. List Results Property-Manager

Figure-110. List Results PropertyManager with expanded Advanced Options rollout

- Select the desired parameters from the **Advanced Options** rollout to set the limits of displayed result.

- Click on the **OK** button from the **PropertyManager**. The **List Results** dialog box will be displayed; refer to Figure-111.

- Click on the **Save** button to save the results in a CSV excel file. Click on the **Close** button to close the dialog box.

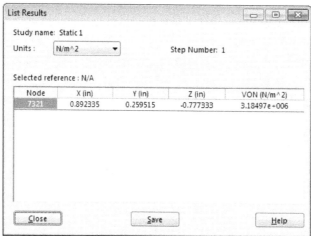

Figure-111. List Results dialog box

List Result Force

This tool is used to display the reaction forces resulting due to the applied load. The steps to use this tool are given next.

- Click on the **List Result Force** tool from the **Results Advisor** drop-down. The **Result Force PropertyManager** will be displayed as shown in Figure-112.
- Select the desired radio button from the **PropertyManager**.
- The blue selection box is activated by default. You need to select the face/edge/point at which you want to check the reaction forces.
- Select the desired unit system from the **Unit** drop-down in the **PropertyManager**.
- Click on the **Update** button from the **PropertyManager** to display the reaction forces. The box containing values of reaction forces will be displayed with the model; refer to Figure-113.

Figure-112. Result Force PropertyManager

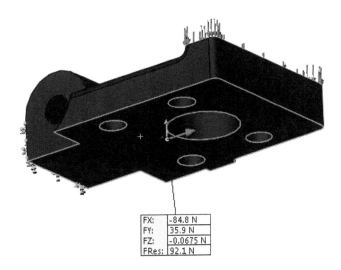

FX:	-84.8 N
FY:	35.9 N
FZ:	-0.0675 N
FRes:	92.1 N

Figure-113. Preview of result force

- Select the **Display resultant forces** check box from **Display Options** section to convert the three force vectors(X,Y, and Z) to a resultant force vector.
- Click on the **Save to file** button from **Report Options** section to save a file of current study.
- After specifying the parameters, click on the **OK** button from the **PropertyManager** to exit the tool.

If you have applied contact forces or Pin/Bolt/Bearing forces, then click on the respective tool from the **Results Advisor** drop-down. The result will be displayed in the same way as discussed earlier.

We will learn about more options related to results while performing the analyses in subsequent chapters.

SELF-ASSESSMENT

Q.1 We can perform static analysis on objects without applying material to object. (T/F)

Q2. If you do not apply fixtures then the object will be free to move in the direction of forces applied and no stress or strain will be induced in it. (T/F)

Q3. Using the Fixed Geometry fixture type, you can fix the face, edge, or point of the object. (T/F)

Q4. The **Fixed Hinge** tool is used to fix a face in such a way that it can slide/roll in its plane but cannot move perpendicular to the plane.

Q5. A live example of elastic support fixture can be an object placed on wood, rubber or any spring base. (T/F)

Q6. Which of the following fixture option makes the object move in specified direction and restricts motion in other directions?

a. Use Reference Geometry
b. Cyclic Symmetry
c. Elastic Support
d. Roller/Slider

Q7. Which of the following rollout in **Force/Torque PropertyManager** is used to define equation for non-uniform force distribution?

a. Force/Torque
b. Nonuniform Distribution
c. Symbol Settings
d. Selection

Q8. .... is product of force and distance.

Q9. .... is force applied per unit area.

Q10. .... force is a force which causes the rotating objects to move outward of their circular path.

FOR STUDENT NOTES

Answer to Self-Assessment:
1. F, 2. T, 3. T, 4. F, 5. T, 6. a, 7. b, 8. Torque, 9. Pressure, 10. Centrifugal

Chapter 3

Preparing Model for Analysis

Topics Covered

The major topics covered in this chapter are:

- *Creating Reference Geometry*
- *Simplifying Model for Analysis*
- *Extrude and Revolve Tool*
- *Splitting Faces and Body*
- *Generating Mid Surface*
- *Adding and Subtracting Bodies*
- *Importing Parts*

INTRODUCTION

No Pal! You can not run analysis directly on a model which has fillets and other unnecessary features. The features like fillet, chamfer, holes (which are not playing role in analysis) increase the solution time and error of the analysis. So, we need to de-feature the model before running analysis in SolidWorks Simulation. There is a separate **CommandManager** in SolidWorks to simplify model called **Analysis Preparation CommandManager**; refer to Figure-1.

Figure-1. Analysis Preparation CommandManager

You can perform many other task like finding mid-surface, splitting faces, and creating reference features using the tools in this **CommandManager**. The most commonly used tools in this **CommandManager** are given next.

REFERENCE GEOMETRY

There are various types of reference geometries that can be created in SolidWorks. All the tools to create these reference geometries are available in the **Reference Geometry** drop-down; refer to Figure-2. The tools available in the drop-down are:

- Plane
- Axis
- Coordinate System
- Point
- Center of Mass
- Mate Reference

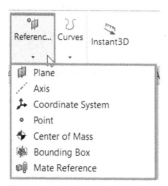

Figure-2. Reference drop-down

These tools are discussed next.

Plane

The **Plane** tool is used to create reference planes. By default, there are three planes available in SolidWorks: **Front**, **Top**, and **Right**. To create more planes follow the steps given next.

- Click on the **Plane** tool from the **Reference Geometry** drop-down. The **Plane PropertyManager** will display as shown in Figure-3.

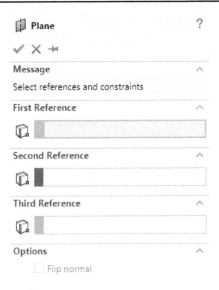

Figure-3. Plane PropertyManager

- You can select maximum three references to create a plane. You can select plane/face, edge/axis/curve, or vertex/point. The ways in which you can create planes by using these references are discussed next.

Creating plane at a distance from plane/face

- To create a plane at a distance from a plane/face, select the plane/face. The updated **Plane PropertyManager** will display as shown in Figure-4.
- Specify the desired distance in the spinner.
- Click on the **OK** button to create the plane.

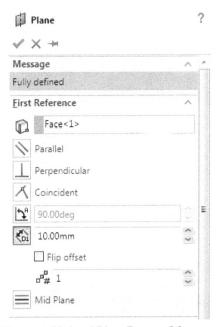

Figure-4. Updated Plane PropertyManager

Creating plane at an angle to plane/face

- Activate the **Plane PropertyManager** and select a plane/face to which you want to specify the angle.
- Click on the **At Angle** button to specify the angle.

• Click in the **Second Reference** box and select the edge or axis to which you want to make the plane coincident or select the two planar point through which you want the plane to pass. Figure-5 shows the plane create by both the methods discussed.

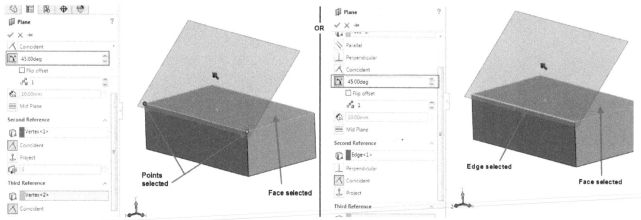

Figure-5. Plane creation at angle

Creating plane passing through points

• Activate the **Plane PropertyManager** and one by one click three points through which you want the plane to pass through. Figure-6 shows the plane passing through three points.

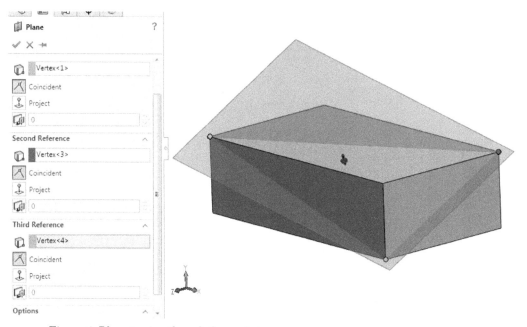

Figure-6. Plane passing through three points

Plane Parallel to Screen

The procedure to create plane parallel to screen is given next.

• Make sure no other tool is selected and then right-click on any face, edge, or vertex of the model. A shortcut menu will be displayed; refer to Figure-7.

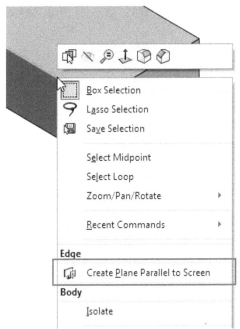

Figure-7. Shortcut menu on right clicking on a face

- Click on the **Create Plane Parallel to Screen** option from the shortcut menu. A plane parallel to screen will be created; refer to Figure-8.

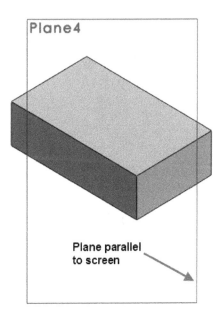

Figure-8. Plane parallel to screen

Axis

The **Axis** tool is used to create reference axes. An axis is useful in creating revolve features or to create planes at angle. The procedure to create axis by using the **Axis** tool is given next.

- Click on the **Axis** tool from the **Reference** drop-down. The **Axis PropertyManager** will display as shown in Figure-9.
- Select the desired button from the **PropertyManager**. The buttons in this **PropertyManager** are explained next.

Figure-9. Axis PropertyManager

One Line/Edge/Axis

Select this button if you want to create an axis coincident to the selected line/edge/axis. After selecting this button, click on the line/edge/axis. The axis will be created coincident to the selected line/edge/axis; refer to Figure-10.

Figure-10. Axis created on edge

Two Planes

Select the **Two Planes** button if you want to create axis at the intersection of the two selected planes/faces. After selecting this button, click on the two intersecting. The axis will be created at the intersection; refer to Figure-11.

Figure-11. Axis at intersection of planes

Two Points/Vertices

Select the **Two Points/Vertices** button if you want to create axis passing through the selected two points/vertices; refer to Figure-12.

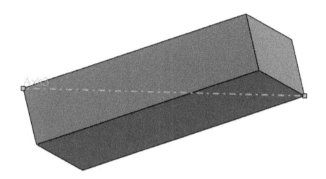

Figure-12. Axis passing through two points or vertices

Cylindrical/Conical Face

Select the **Cylindrical/Conical Face** button and select a cylindrical/conical face. An axis passing through center of cylindrical/conical face will be created; refer to Figure-13.

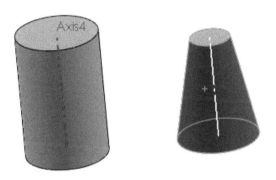

Figure-13. Axis through cylinder and conical

Point and Face/Plane

Select the **Point and Face/Plane** button if you want to create an axis passing through the selected point and perpendicular to the selected face/plane.

Coordinate System

The **Coordinate System** tool is used to create reference coordinate system. The steps to create coordinate system are explained next.

- Click on the **Coordinate System** tool from the **Reference** drop-down. The **Coordinate System PropertyManager** will display as shown in Figure-14.

Figure-14. Coordinate System
PropertyManager

- Click on the point where you want to place the coordinate system.
- Click in the box for which you want to specify direction reference and select the reference like plane, axis and so on. Figure-15 shows a coordinate system created on the face.

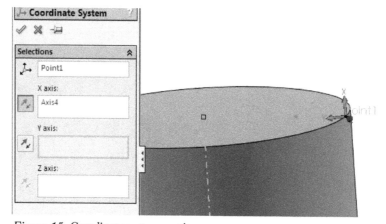

Figure-15. Coordinate system creation

Point

The **Point** tool is used to create reference points on the model. The steps to create points are given next.

- Click on the **Point** tool from the **Reference** drop-down. The **Point PropertyManager** will display as shown in Figure-16.
- Select the desired button to specify the type of point you want to create. In this case, we have selected **Center of Face** button.
- Select the reference (face of the model in this case). Preview of the point will display; refer to Figure-17.

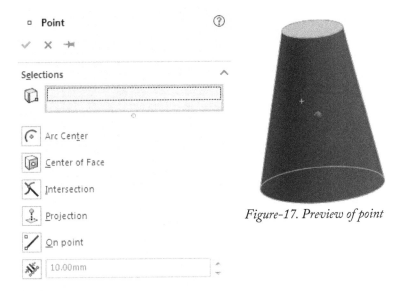

Figure-16. Point PropertyManager

Figure-17. Preview of point

- Click on the **OK** button to create the point.

You can create array of points along a curve by selecting  button.

Center of Mass

The **Center of Mass** tool is used to display the center of mass of the model. The coordinates of center of mass are generally required in some calculations related to inertia of the objects. Identification of center of mass is also helpful in checking the stability of object in constraint free environment. To display the center of mass, click on the **Center of Mass** tool from the **Reference** drop-down and the center of mass will display in the viewport; refer to Figure-18.

Figure-18. Center of mass of cylinder

Bounding Box

The **Bounding Box** tool is used to create an envelope for parts created in the drawing area. The procedure to use this tool is given next.

- Click on the **Bounding Box** tool from the **Reference Geometry** drop-down in the **Features CommandManager** of **Ribbon**. The **Bounding Box PropertyManager** will be displayed; refer to Figure-19.

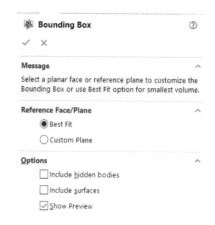

Figure-19. Bounding Box PropertyManager

- Select the **Best Fit** radio button if you want to create the bounding box automatically based on geometry of the part. Select the **Custom Plane** radio button if you want to create bounding box based on selected plane/face. The selection box for plane will be displayed. Click in the selection box and select desired face/plane to be used as boundary reference for bounding box.
- Select the **Include hidden bodies** check box to include hidden bodies of part while creating the bounding box.
- Select the **Include surfaces** check box to include surfaces while creating the bounding box.
- Select the **Show Preview** check box to display preview while create the box.
- After setting desired parameters, click on the **OK** button from the **PropertyManager**.

SIMPLIFY

The **Simplify** tool is used to remove the features that are of no importance in the current analysis. This tool is active in the Ribbon only if there is at least one hole, fillet, chamfer feature on the solid base. The procedure to use this tool is given next.

- Click on the **Simplify** tool from the **Analysis Preparation CommandManager**. The **Simplify** task pane will be displayed at the right in the application window; refer to Figure-20.

Figure-20. Simplify taskpane

- Click in the **Features** drop-down of task pane and select the check boxes of features that you want to be simplified.
- Specify the desired simplification factor in the **Simplification factor** spinner and select the radio button below it to set the simplification factor type.
- Click on the **Find Now** button to search for the features that can be simplified based on your inputs. The features that can be simplified or say suppressed are displayed in the **Results** area of the task pane.
- Select the features that do not have any effect on your analysis while holding CTRL key; refer to Figure-21.

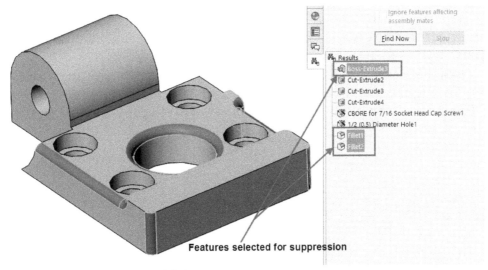

Figure-21. Features to be simplified

- Specify the name of new configuration of part/assembly to be created with suppressed features in the **Name** edit box at the bottom of the task pane. Make sure the **Create derived configuration** check box is selected.
- Click on the **Suppress** button at the bottom in the task pane. A new configuration with suppressed features will be created on which you can perform the analysis; refer to Figure-22.

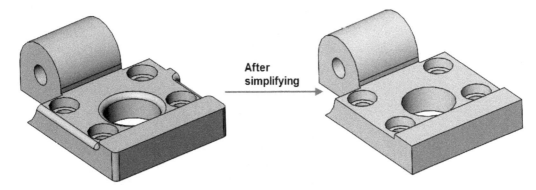

Figure-22. Simplifying model

EXTRUDED BOSS/BASE TOOL

Extruded Boss/Base tool is used to create a solid volume by adding height to the selected sketch. In other words, this tool adds material (by using the boundaries of sketch) in the direction perpendicular to the plane of sketch. In the term Boss/Base; the Base denotes the first feature and Boss denotes the feature created on any other feature. The steps to create extruded feature is given next.

• Click on the **Extruded Boss/Base** tool from the **Ribbon**. The **Extrude PropertyManager** will display; refer to Figure-23.

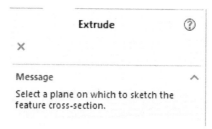

Figure-23. Extrude PropertyManager

• Select a plane from the planes displayed in the viewport. The sketching environment will display with the sketch tools activated.
• Create a closed sketch and then click on the **Exit Sketch** button from the viewport as shown in Figure-24. You can also select the **Exit Sketch** button from the **Ribbon**. The **Boss-Extrude PropertyManager** will display as shown in Figure-25.

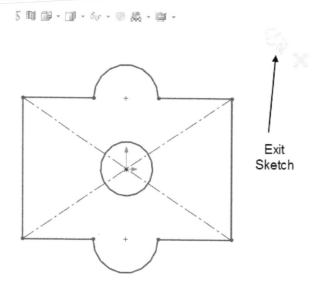

Figure-24. Sketch environment of extrude

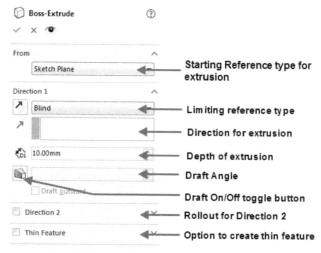

Figure-25. Boss-Extrude PropertyManager

- Click on the **Starting reference** drop-down and select the desired option.

There are four options in this drop-down; **Sketch Plane**, **Surface/Face/Plane**, **Vertex**, and **Offset**. The **Sketch Plane** is selected by default. Select this option if you want the extrusion to start from sketching plane. Select the **Surface/Face/Plane** option to start the extrusion from the selected surface/face/plane. Select the **Vertex** option to start extrusion from selected vertex. Select the **Offset** option if you want to start at specified distance from the sketching plane; refer to Figure-26.

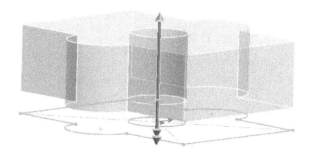

Figure-26. Preview of offset option

- Click in the **Limiting reference type** drop-down and select the reference for end of extrusion.

There are six options in the drop-down; **Blind, Up To Vertex, Up To Surface, Offset From Surface, Up To Body**, and **Mid Plane**. If you have selected **Blind** or **Mid Plane** option, then you need to specify the distance value in the **Height of extrusion** spinner. If you have selected any of the other option then select the respective reference from the viewport. Figure-27 shows preview of extrusion by using the **Mid Plane** option. Note that if **Mid Plane** option is selected then the Direction 2 rollout will not display.

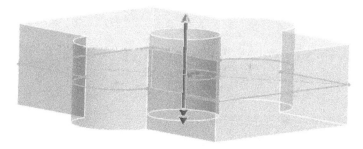

Figure-27. Mid-Plane extrusion

- Click in the **Direction of extrusion** selection box and select the reference if you do not want to extrude perpendicular to the sketching plane and want to extrude along selected axis/plane
- Click in the edit box for extrusion height and enter the desired extrusion height or you can set the value by using spinner.
- Click on the **Draft On/Off** button to apply draft angle on the vertical faces of the model. On selecting this button, **1°** draft will be applied by default taking the sketching plane as reference. Select the **Draft outward** check box to apply draft angle outwards on the vertical faces of extrusion. Specify the draft angle in the **Draft Angle** spinner.

The parameters you specified above can also be applied to the opposite direction. To apply these parameters, select the **Direction 2** check box. The parameters for the opposite direction will display.

- Select the **Thin Feature** check box to create the thin walled extrusion. Enter the thickness in the **Thickness** edit box of the **Thin Feature** rollout. Figure-28 shows a thin featured extrusion. **Note that if open sketch is selected for extrude then this option gets selected automatically.**

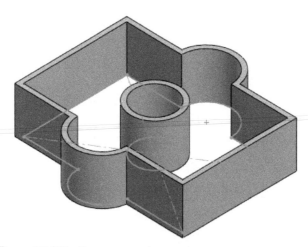

Figure-28. Thin Feature extrusion

- If you want to close the start and end face of the extrusion then select the **Cap Ends** check box; refer to Figure-29.

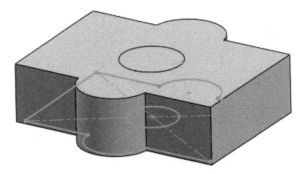

Figure-29. Extrusion with cap ends

- Once you have finished creating the feature, click on the **Detailed Preview** button from the **PropertyManager** to verify the feature; refer to Figure-30.

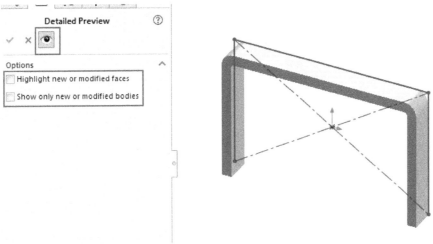

Figure-30. Detailed Preview button

- Select the **Highlight new or modified faces** check box if you want to highlight the new/modified faces only. Similarly, you can select the **Show only new or modified bodies** check box if you want to display only new or modified objects.

You can also activate the **Extrude Boss/Base** tool by ALT key. To do so, press the ALT key from keyboard and click on the shaded sketch section in the Sketching environment. The **Extrude** button will be displayed; refer to Figure-31. Click on the **Extrude** button to display **Extrude PropertyManager**. Rest of the procedure is same as discussed earlier.

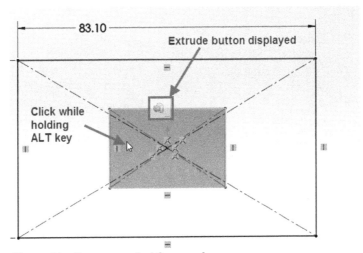

Figure-31. Alternate method for extrude

REVOLVED BOSS/BASE TOOL

Revolved Boss/Base tool is used to create a solid volume by revolving a sketch about selected axis. In other words, if you revolve a sketch about an axis then the volume that is swept by revolved sketch boundary is called revolved boss/base feature. The steps to create revolved boss/base feature are given next.

- Click on the **Revolved Boss/Base** tool. If you have not selected any existing sketch, then the **Revolve PropertyManager** displays as shown in Figure-32.

Figure-32. Revolve PropertyManager

- Select a plane if you want to create a new sketch or select an already created sketch. In our case, we are selecting an already created sketch.
- Select the region of the sketch that you want to revolve if you have multiple loops in sketch; refer to Figure-33. The updated **Revolve ProperyManager** will display as shown in Figure-34.

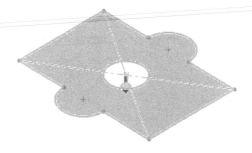

Figure-33. Region selected for revolve

Figure-34. Updated Revolve Property Manager

- Click in the **Axis of Revolution** selection box to select the axis. Select the edge, line, or center line about which you want to revolve the sketch. Preview of the revolve feature will display; refer to Figure-35.

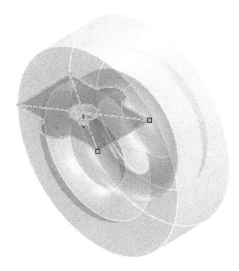

Figure-35. Preview of revolve feature

- Click on the **Revolve Type** drop-down and specify the revolution limiting reference. The options in this drop-down are same as discussed for **Extruded Boss/Base** tool.
- If you have selected **Blind** option in the **Revolve Type** drop-down then specify the degrees of revolution by using the **Angle** spinner.
- Click on the **Direction 2** check box to revolve in the direction opposite the earlier on. The options in the **Direction 2** rollout are same as discussed earlier.
- You can also create thin feature by selecting the **Thin Feature** check box. **Note that if you select an open sketch then this option is automatically selected**.
- Click in the **Selected Contours** box to add more sketches for revolution and select the sketches you want to revolve.

Figure-36 shows a sketch, axis of revolution, and resulting revolve feature preview.

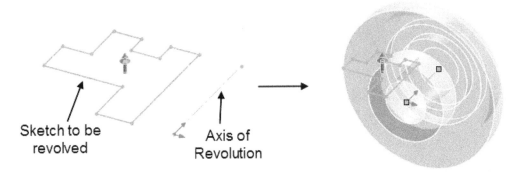

Sketch to be revolved   Axis of Revolution

Figure-36. Revolved feature

EXTRUDED CUT

The **Extrude Cut** tool is used to remove material by extruding the sketch. The steps to use this tool are given next.

- Click on the **Extruded Cut** tool. The **Cut-Extrude PropertyManager** will display.
- Select the face from which you want to start removing the material. The sketch environment will activate.
- Click the sketch using which you want to remove the material.
- Click on the **Exit Sketch** button from the **Ribbon**. The preview of cut feature will display; refer to Figure-37.

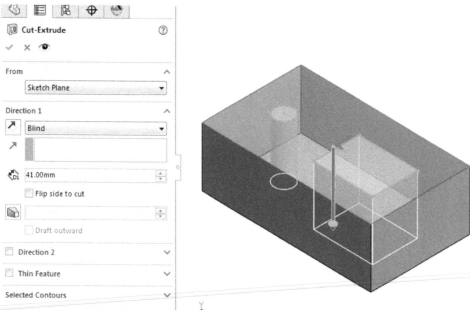

Figure-37. Preview of extrude cut feature

- Specify the depth of material removal and other parameters in the **PropertyManager** as discussed for **Extrude Boss/Base** tool and then click on the **OK** button.

MIRROR

The **Mirror** tool is used to create mirror copy of the features in the Modeling environment. The procedure to create mirror is given next.

- Select all the features that you want to mirror.
- Click on the **Mirror** tool from the **Analysis Preparation CommandManager** in the **Ribbon**. The **Mirror PropertyManager** will display as shown in Figure-38.
- Select the plane or face about which you want to mirror the features. Preview of the mirror will be displayed; refer to Figure-39.
- Click on the **OK** button from the **PropertyManager** to create the feature.

Figure-38. Mirror PropertyManager

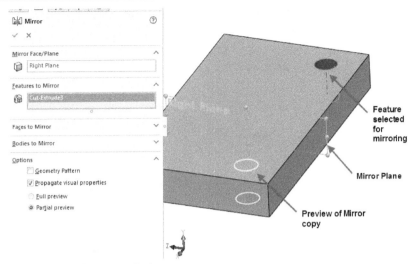

Figure-39. Preview of mirror

If you want to mirror a complete body with respect to mirror plane as in the above figure, then follow the steps given next.

- Click on the **Mirror** tool. The **Mirror PropertyManager** will display.
- Expand the **Bodies to Mirror** rollout and click on the bodies in the viewport that you want to mirror.
- Click in the **Mirror Face/Plane** box in the **Mirror Face/Plane** rollout of the **ProperyManager** and select the mirror plane. Preview of mirror will display; refer to Figure-40. Make sure that you clear the **Merge solids** check box before creating the mirror copy.
- Click on the **OK** button to create the mirror copy.

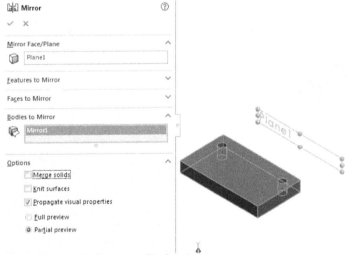

Figure-40. Preview of body mirror

SPLIT LINE TOOL

In some of the cases, you need to apply load or constraint at a confined area of face rather then the full surface or area. In that case, you need to split the face. The steps given next can explain the procedure of splitting.

- Click on the **Split Line** tool from the **Ribbon**. The **Split Line PropertyManager** will be displayed as shown in Figure-41.

Figure-41. Split Line PropertyManager

- By default, the **Silhouette** radio button is selected. You are supposed to select the direction of force and faces to split.
- Click in the blue selection box and select the faces to split.
- Click in the pink selection box and select the direction of split. In this you can also select a plane to specify the direction.
- Click in the **Reverse** direction button to reverse the selected direction.
- If you want to split the body or face at a particular angle, click in the **Angle** edit box and specify the desired value.
- After specifying the parameters, click on the **OK** button from **Spit Line PropertyManager**. The model will be split at the selected reference; refer to Figure-42.

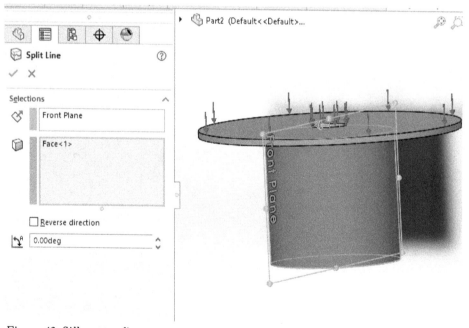

Figure-42. Silhouette split

- Select the **Projection** radio button from **Type of Split** section and you are supposed to select a sketch for dividing the face. Figure-43 shows a part that need to be divided from mid.

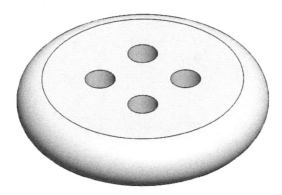

Figure-43. Part to be splitted

- A line sketch is drawn at the center of the mold part on Right-plane so that it can divide the part into two pieces; refer to Figure-44.

Note that to draw the sketch, you need to cancel the **Split Line** tool, select the **Sketch CommandManager** from the **Ribbon** and after you draw the sketch, you need to select the **Split Line** tool again.

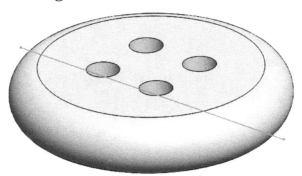

Figure-44. Sketched line drawn at the center

- Select the line sketch, it will be displayed in the first selection box in the **Split Line PropertyManager**.
- Click in the next selection box. You will be asked to select the faces to be split up.
- Select all the faces of the part that you want to split by using the projection of sketch; refer to Figure-45.
- Select the **OK** button from the **PropertyManager**, the part will be displayed as shown in Figure-46.

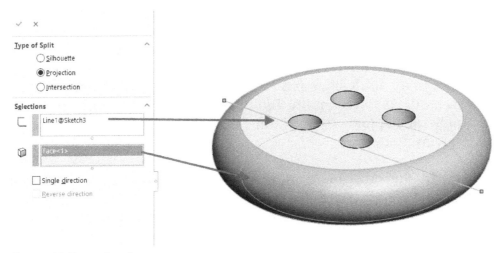

Figure-45. Faces selected

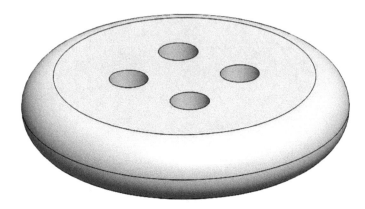

Figure-46. Part after splitting

- In the same way, you can split the part by using the **Intersection** radio button; refer to Figure-47.

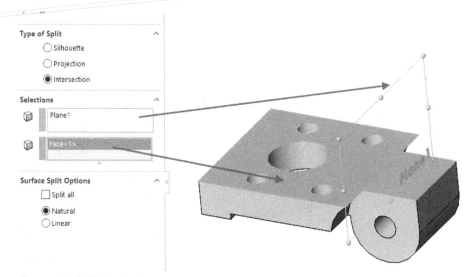

Figure-47. Splitting by intersection

MID SURFACE

The **Mid Surface** tool is used to create a surface at the mid of selected faces of the part. You may ask what is the benefit of this for simulation. Most of the time when you remove extra features from your model, the model is a plane piece of metal with different thicknesses at different places. When you perform analysis on a solid, the number of elements during meshing are very high due to stacking of elements along Z direction apart from X and Y directions. If you create mid surface of solid and apply thickness using the **Shell Manager** then number of elements are reduced. In turn, the time taken by analysis is reduced. When you apply thickness using **Shell Manager** then 3D model of part remains in surface but while solving analysis equation, value of thickness is applied in equations automatically. The procedure to create Mid surface is given next.

- Click on the **Mid Surface** tool from the **Analysis Preparation CommandManager**. The **MidSurface PropertyManager** will be displayed; refer to Figure-48. Also, you will be asked to select the first face for creating mid-surface.

Figure-48. MidSurface PropertyManager

- Select the first face of the model. You will be asked to select the second face of the model in current set.
- Select the second face, a set will be created in the **Face pairs** selection box; refer to Figure-49.

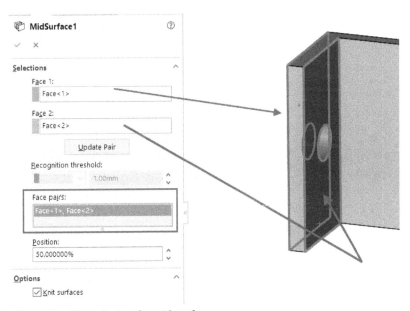

Figure-49. Face selection for mid surface

- After selection the **Face 1** and **Face 2** selection boxes will become empty prompting you to select the next set.
- If you want to find the face pairs automatically for mid surface then do not select any face manually. In the **Recognition Threshold** drop-down, select the desired operator and then specify the value of threshold in the edit box next to it. Click on the **Find Face Pairs** button from the **PropertyManager**. Most of the faces will get selected automatically; refer to Figure-50.

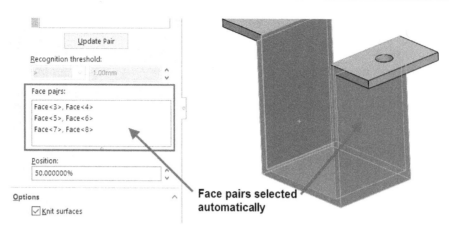

Figure-50. Face pairs selected automatically

- Select the rest of the face as discussed earlier if required.
- Set the position of the mid-surface using the **Position** spinner. If the position value is **50%** then mid surface will be created at the center of the face pairs.
- Click on the **OK** button from the **PropertyManager** to create the surface. Hide rest of the feature to check the mid-surface; refer to Figure-51.

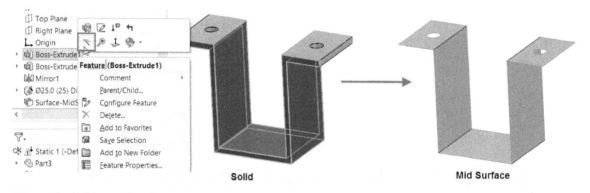

Figure-51. Solid to mid surface

There are various surface designing tools in **Surfaces** drop-down which are not in the scope of this book. You can refer to our book SolidWorks 2020 Black Book for these tools.

SPLIT TOOL

The **Split** tool is used to create multiple bodies of a solid body by using a trimming tool. The procedure to use this tool is given next.

- Click on the **Split** tool from the **Analysis Preparation CommandManager** in the **Ribbon**. The **Split PropertyManager** will be displayed; refer to Figure-52.

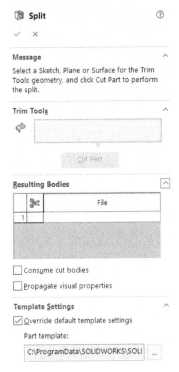

Figure-52. Split PropertyManager

- Select the plane/surface/sketch to be used as trimming tool. Note that the plane/surface should intersect the part.
- Click on the **Cut Part** tool from the **PropertyManager**. The part will be separated into multiple bodies and displayed in **Resulting Bodies** section.
- Select the required check boxes of specific bodies from the **Resulting Bodies** rollout to create splits.
- Select the **Consume cut bodies** check box to consume the body selected in the **Resulting Bodies** rollout.
- Select the **Propagate visual properties** check box to apply the same properties to the split bodies which were applied to the main body.
- Select the Override default template settings check box from Template Settings section to override the default template settings of the current tool.
- After specifying the parameters, click on the **OK** button from the **Split PropertyManager**. The **Split** feature will be added in **Design Tree**.

COMBINE TOOL

The **Combine** tool is used to add/subtract two or more solid bodies. The procedure to use this tool is given next.

- Click on the **Combine** tool from the **Analysis Preparation CommandManager**. The **Combine PropertyManager** will be displayed; refer to Figure-53.

Figure-53. Combine PropertyManager

Adding Solid Bodies

- Select the **Add** radio button from the **Operation Type** rollout and select all the bodies that you want to be combined.
- Click on the **OK** button from the **PropertyManager** to combine bodies.

Subtracting Solid Bodies

- Select the **Subtract** radio button from the **Operation Type** rollout and select the main body from which you want to subtract other solid bodies. You will be asked to select the bodies to be subtracted.
- Select the bodies that you want to be subtracted from the main body.
- Click on the **OK** button from **Combine PropertyManager** to subtract the selected body.

INTERSECT TOOL

The **Intersect** tool is used to intersect solids, surfaces, or planes to modify existing geometry or create a new geometry of intersection boundaries. The procedure to use this tool is given next.

- Click on the **Intersect** tool from the **Analysis Preparation CommandManager**. The **Intersect PropertyManager** will be displayed; refer to Figure-54. You will be asked to select the bodies which are intersecting.

Figure-54. Intersect Property Manager

- Select the intersecting solids, surfaces, or planes which are forming closed section.
- Select the desired radio button from the **PropertyManager** to create desired regions after intersection.
- Click on the **Intersect** button from the **PropertyManager**. Various regions of the solid will be displayed in the drawing area as well as in the **Region to Exclude** rollout; refer to Figure-55.

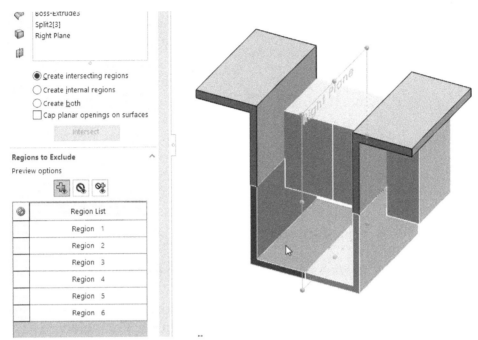

Figure-55. Regions created from solid

- Select the check box from the **Region to Exclude** rollout to exclude selected region.
- If you want to merge all the created regions to form one body then select the **Merge result** check box from **Options** rollout of the **PropertyManager**.
- Similarly, select the **Consume surfaces** check box to delete the interesting surfaces after creating the regions.
- After specifying the parameters, click on the **OK** button from the **PropertyManager** to create the body/regions. The **Intersect** feature will be added in **Design Tree**.

MOVE/COPY BODIES TOOL

The **Move/Copy Bodies** tool is used to move or copy selected bodies. The procedure to use this tool is given next.

- Click on the **Move/Copy Bodies** tool from the **Analysis Preparation CommandManager**. The **Move/Copy Body PropertyManager** will be displayed as shown in Figure-56.

Figure-56. Move/Copy Body PropertyManager

- The **Bodies to Move** selection box is active by default. You need to click on the body or a part to select.
- Click in the **Entities to Mate** selection box from **Mate Settings** section to select two entities to mate together.
- After selecting the two entities, click on the required button of mating type.
- Click on the **Add** button from **Mate Settings** section The selected entities will be mate and displayed in **Mates** box.
- Click on the **Translate/Rotate** button from **Options** section to move, copy, rotate the body by specifying parameters; refer to Figure-57.

Figure-57. Parameters of Translate button

- Click in the specific edit box from **Translate** section and enter the desired value.
- Clicking the **Constraint** button from **Rotate** section will return you back to the initial screen **PropertyManager**.
- After specifying the parameters, click on the **OK** button from **Move/Copy Body PropertyManager**.

Similarly, you can use the **Delete/Keep Body** tool and other tools of SolidWorks.

IMPORTING PART

- Click on the **Open** button from the **Quick Access Toolbar** or press CTRL+O from the keyboard. The **Open** dialog box will be displayed.
- Double-click on the desired file that you want to open. Make sure the **All Files** or desired file type is selected in the **File Type** drop-down in the dialog box; refer to Figure-58. The respective **SOLIDWORKS Converter** dialog box will be displayed; refer to Figure-59.

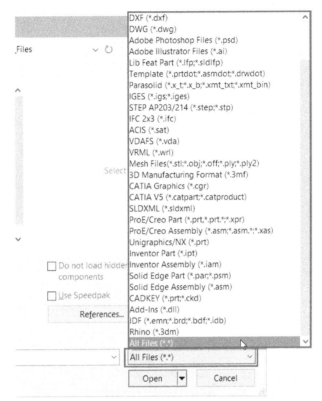

Figure-58. File Type drop-down

Figure-59. SOLIDWORKS Converter dialog box

- Select the **Import geometry directly** radio button and select the desired radio button to import the geometries of selected file. If you select the **BREP** radio button the model is imported as solid with Boundary Representation data. If you select the Knitting radio button then the model will be imported as surfaces knitted together and if you also select the **Try forming solid model(s)** check box then system will form a solid if surface form closed boundaries. If you select the Do not knit the surfaces will be imported in SolidWorks.

- If you want SolidWorks to recognise as many features as possible in the imported model then select the **Analyze the model completely** radio button from the dialog box.

- Select the desired check boxes from the dialog box to import respective parameters of the model.

- Click on the **OK** button from the dialog box. If you have selected the **Analyze the model completely** radio button earlier then an information box will be displayed; refer to Figure-60. Click on the **Body** button and select the desired check boxes. The **SOLIDWORKS** dialog box will be displayed as shown in Figure-61.

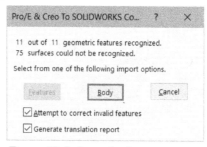

Figure-60. Information box

Figure-61. SOLIDWORKS dialog box

- Click on the **Yes** button from the dialog box to start import diagnostic which help in finding any inconsistency of imported model. The **Import Diagnostics PropertyManager** will be displayed; refer to Figure-62.

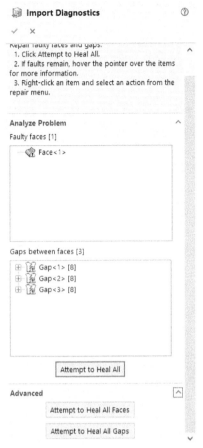

*Figure-62. Import Diagnostics Prop-
ertyManager*

- Click on the **Attempt to Heal All** button to heal the faces/gaps. The faces/gaps which are not able to heal automatically should either be deleted or remodelled.
- Click on the **OK** button from the **PropertyManager** to solve problems of imported parts.

FOR STUDENT NOTES

Chapter 4

Static Analysis

Topics Covered

The major topics covered in this chapter are:

- *Starting Static Analysis.*
- *Applying Material.*
- *Defining Fixtures*
- *Applying loads*
- *Defining Connections*
- *Simulating Analysis.*
- *Interpreting results*

LINEAR STATIC ANALYSIS

Linear static analysis is performed to calculate stresses, displacements, strains, reaction forces, and error estimates under various loading conditions. On applying the loads, the body gets deformed and the loads are transmitted throughout the body. The deformation and other effects of body are studied under this analysis.

Assumptions

Some assumptions are made in this type of analysis, like:

1. The loads applied do not vary with time.
2. All loads are applied slowly and gradually until they reach to the full magnitude and after reaching the full magnitude, the loads remain constant. Thereby, neglecting impact, inertial, and damping forces.
3. The materials applied to the components satisfy the Hooke's law.
4. The change in stiffness due to loading is neglected.
5. Boundary conditions do not vary during the application of loads. Loads must be constant in magnitude, direction, and distribution.

Geometry Assumptions

1. The part model must represent the required CAD geometry.
2. Only the internal fillets in the area of interest will be included in the study.
3. Shells are created when thickness of the part is small in comparison to its width and length.
4. Thickness of the shell is assumed to be constant.
5. If the dimensions of a particular part are not critical and do not affect the analysis results, some approximations can be made in modeling the particular part.
6. Primary members of structure are long and thin like a beam then idealization is required.
7. Local behavior at the joints of beams or other discontinuities are not of primary interest so no special modeling of these area is required.
8. Decorative or external features will be assumed insignificant to the stiffness and the performance of the part and will be omitted from the model.

Material Assumptions

1. Material remain in the linear regime. It is understood that either stress levels exceeding yield or excessive displacements will constitute a component failure. That is non linear behavior cannot be accepted.
2. Nominal material properties adequately represent the physical system.
3. Material properties are not affected by load rate.
4. Material properties can be assumed isotropic (Orthotropic) and homogeneous.
5. Part is free of voids or surface imperfections that can produce stress risers and skew local results.
6. Actual non linear behavior of the system can be extrapolated from the linear material results.
7. Weld material and the heat affected zone will be assumed to have same material properties as the base material.
8. Temperature variations may have a significant impact on the properties of the materials used. Change in material properties is neglected.

Boundary Conditions Assumptions

1. Choosing proper BC's require experience.
2. Using BC's to represent parts and effects that are not or cannot be modeled leads to the assumption that the effects of these un-modeled entities can truly be simulated or has no effect on the model being analyzed.
3. For a given situation, there would be many ways of applying boundary conditions. But these various alternatives can be wrong if the user does not understand the assumptions they represent.
4. Symmetry/ anti-symmetry/ reflective symmetry/ cyclic symmetry conditions if exists can be used to minimize the model size and complexity.
5. Displacements may be lower than they would be if the boundary conditions being more appropriate. Stress magnitudes may be higher or lower depending on the constraint used.

Fasteners Assumptions

1. Residual stress due to fabrication, pre-loading of bolts, welding and/or other manufacturing or assembly processes are neglected.
2. Bolt loading is primarily axial in nature.
3. Bolt head or washer surface torque loading is primarily axial in nature.
4. Surface torque loading due to friction will produce only local effects.
5. Bolts, spot welds, welds, rivets, and/or fasteners which connect two components are considered perfect and acts as rigid joint.
6. Stress relaxation of fasteners or other assembly components will not be considered. Load on threaded portion of the part is evenly distributed on engaged threads
7. Failure of fasteners will not be reflected in the analysis.

General Assumptions

1. If the results in the particular area are of interest, then mesh convergence will be limited to this area.
2. No slippage between interfacing components will be assumed.
3. Any sliding contact interfaces will be assumed frictionless.
4. System damping will be normally small and assumed constant across all frequencies of interest unless otherwise available from published literature or actual tests.
5. Stiffness of bearings in radial or axial directions will be considered infinite.
6. Elements with poor or less than optimal geometry are only allowed in areas that are not of concern and do not affect the overall performance of the model.

When a system under load is analyzed with linear static analysis, the linear finite element equilibrium equations are solved to calculate the displacement components at all nodes. These results are used to calculate the strained components. These strain results along with stress-strain relationship helps to calculate stresses.

The procedure given next is generic and can be applied on any model and assembly. But, we will use a model to help you understand the complete procedure with real world problems.

PERFORMING STATIC ANALYSIS

Problem:

We have a component which is fixed by its round boss feature and load of 980 N (100 kgf approx) is applied on the its flat top; refer to Figure-1. We need to find out the Factor of Safety of the model and make it optimized.

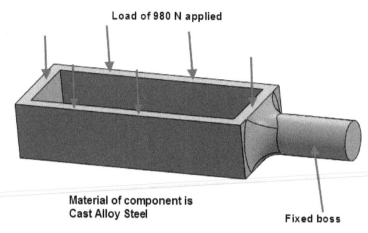

Figure-1. Problem of static analysis

To perform the static analysis on any model, make sure that you have opened the model file in SolidWorks. The file used in this example can be found in the resource kit for this book. After opening the file follow the steps given next.

Starting the Static Analysis

- Open the part on which you want to perform the analysis.
- Click on the **SolidWorks Simulation** button from the **SOLIDWORKS Add-Ins** tab of the **Ribbon** to add **Simulation** tab in the **Ribbon**, if not added already.
- Click on the **New Study** button from **Simulation** tab. The **Study PropertyManager** will be displayed; refer to Figure-2. In this **Study PropertyManager**, list of analysis studies that can be performed, will be displayed.

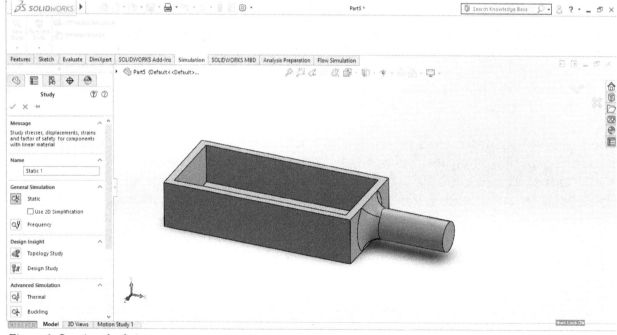

Figure-2. Starting Analysis

- Click on the **Static** button if not selected. Specify the name of the analysis in **Study name** edit box displayed above the list of analyses.
- Click on the **OK** button from the **Study PropertyManager**. **Static 1** tab will be added at the bottom of the modeling area in the **StudyBar**. Also, the tools related to the analysis will be displayed in the **Ribbon**; refer to Figure-3.

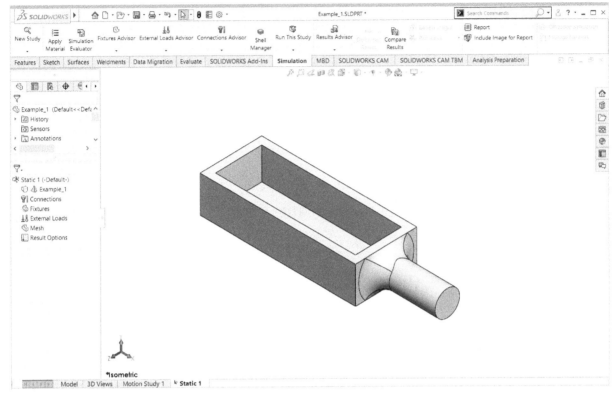

Figure-3. Analysis environment

Applying Material

- Click on the **Apply Material** button from the **Ribbon**. The **Material** dialog box will be displayed; refer to Figure-4.

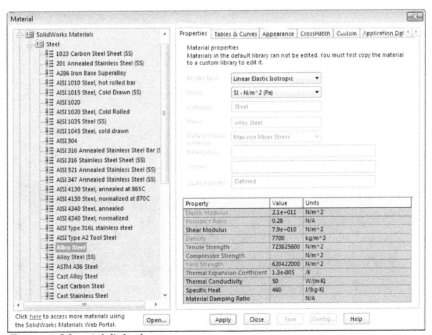

Figure-4. Material dialog box

- Select the **Cast Alloy Steel** from the left area of the dialog box and select the **Apply** button to apply the material.
- Click on the **Close** button to close the dialog box.

Applying Fixtures

- Click on the down button below **Fixtures Advisor** drop-down from the **Ribbon**. The tools related to fixtures will be displayed; refer to Figure-5.

Figure-5. Fixtures Advisor

- Click on the **Fixed Geometry** button from the tool list. The **Fixture PropertyManager** will be displayed; refer to Figure-6.

Figure-6. Fixture Property Manager

- Select the round face of the model to make it fixed; refer to Figure-7.

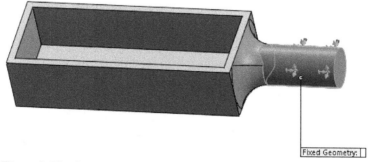

Figure-7. Fixed geometry

- Click on the **OK** button from the **PropertyManager** to fix the round face.

Now, we need to specify the loads that are being applied on the component.

Applying Loads

- Click on the down arrow below **External Loads Advisor** button in the **Ribbon**. List of the tools related to loads will be displayed; refer to Figure-8.

Figure-8. External loads advisor

- Click on the **Force** button from the list. The **Force/Torque PropertyManager** will be displayed as shown in Figure-9.

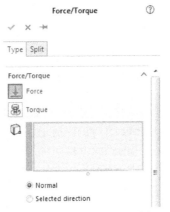

Figure-9. Force Torque PropertyManager

- Select the face as shown in Figure-10 to apply the force load.

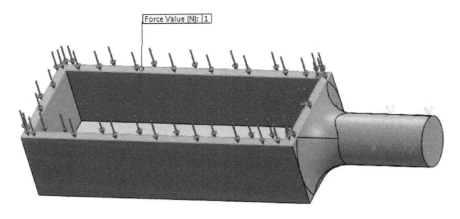

Figure-10. Face selected for applying load

- Click in the **Force Value** box in the **PropertyManager** or in the modeling area and enter the value as **980** which is equal approximately to **100** Kilograms of weight. Refer to Figure-11. Note that you can change the unit of force from the **Unit** drop-down to enter 100 kgf or 220 lbf.

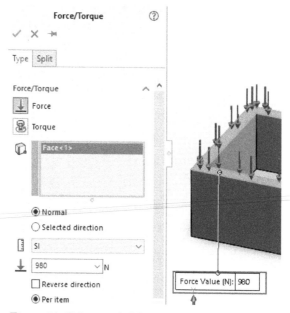

Figure-11. Value specified for load

- Click on the **OK** button from the **Force/Torque PropertyManager**. The force will be applied on the selected face.

For this particular problem, we do not require any connection to be applied since this is not an assembly. Now, we are ready to divide the model into small elements with the help of meshing tools.

Meshing the model

Note that you can skip this step if you do not want to explicitly define the element size for mesh. While running the analysis, system will automatically create the suitable mesh for you. To explicitly define the size of mesh elements, follow the steps given next.

- Click on the down arrow below the **Run This Study** button in the **Ribbon**. The list of tools will be displayed; refer to Figure-12.

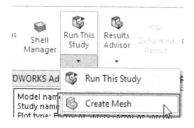

Figure-12. Create Mesh tool

- Click on the **Create Mesh** tool from the list. The **Mesh PropertyManager** will be displayed; refer to Figure-13.
- Click on the **Mesh Parameters** check box. The rollout will expand and the options related to element size and tolerance will be displayed; refer to Figure-14.

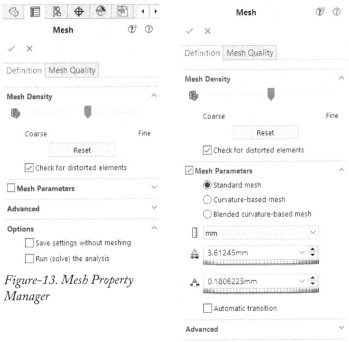

Figure-13. Mesh Property Manager

Figure-14. Mesh Parameters

- Set the desired values in the parameter boxes of the rollout and click on the **OK** button from the **PropertyManager**. Note that the smaller value given in the boxes will make the analysis slower because of more calculations. The meshed model will be displayed as in Figure-15.

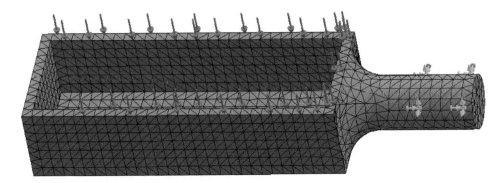

Figure-15. Meshed model

- To change the meshing for a particular area of the model, right-click on the **Mesh** node in the **Analysis Manager** displayed at the left of the modeling area. A shortcut menu will be displayed; refer to Figure-16.
- Click on the **Apply Mesh Control** button. The **Mesh Control PropertyManager** will be displayed as shown in Figure-17.

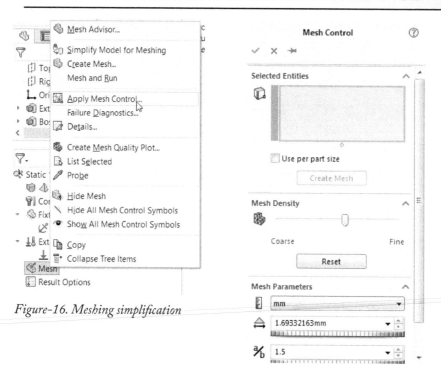

Figure-16. Meshing simplification

Figure-17. Mesh Control PropertyManager

- Select the faces of the model for which you want to change the mesh elements sizes; refer to Figure-18.

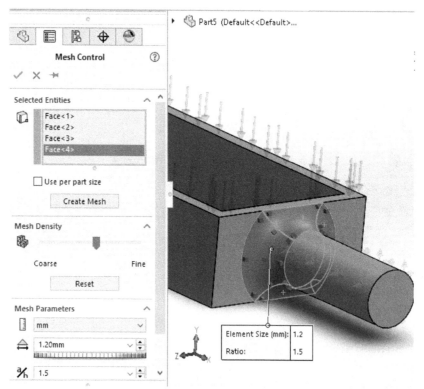

Figure-18. Faces for changing mesh size

- Move the slider in the **Mesh Density** rollout towards coarse or fine as per your requirement. In this case, the selected area will be facing more stress and we are concerned more about this area. So, we will make fine mesh at the selected faces with element size of **1.2** mm.

- Click on the **OK** button from the **Mesh Control PropertyManager** to create the updated mesh. Note that increasing or decreasing the element size will also affect the accuracy of results. Decreasing the element size gives more accurate results but take more processing.

Running the Analysis

- Click on the **Run This Study** button from the **Ribbon**. The system will start solving the analysis.
- After the completion of process, the result of analysis will be displayed; refer to Figure-19.

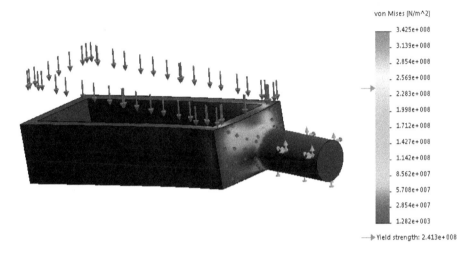

Figure-19. Analysis result

- By default, results related to stress (vonMises) are displayed in the modeling area.
- To check **Displacement** and **Strain** related results, double-click on the respective option from the **Analysis Manager** in the left.

Results (Factor of Safety)

- To check results other than the default displayed, right-click on the **Results** node from the **Analysis Manager**. A shortcut menu will be displayed; refer to Figure-20.
- Click on the **Define Factor Of Safety Plot** option to display the results related to **Factor Of Safety**. The **Factor of Safety PropertyManager** will be displayed as shown in Figure-21.

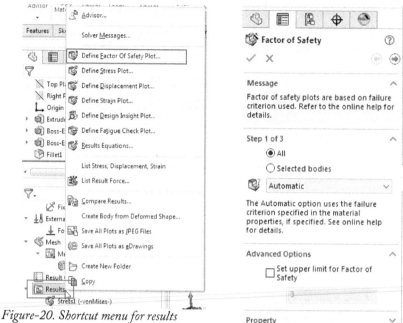

Figure-20. Shortcut menu for results

Figure-21. Factor of Safety
PropertyManager

- If you have a special criteria for checking safety factor then select the respective option from the **Criterion** drop-down in the **Step 1 of 3** rollout otherwise system will check the material properties and automatically decide the best criteria.
- Click on the **OK** button from the **PropertyManager**. The **Factor of Safety plot** will be displayed in the modeling area; refer to Figure-22.

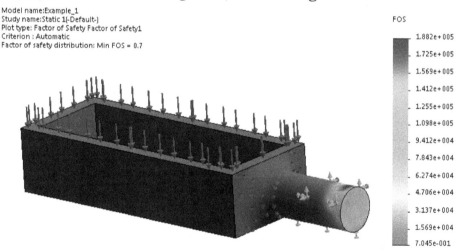

Figure-22. Factor of Safety plot

Optimizing product

- If you check in the results displayed above the model, the **Factor of safety distribution Min FOS = 0.7** is displayed which shows that the model is not safe at the locations displayed in dark red color. Note that for object to be safe for loading, **the FOS should be above 1**.
- Here, we need to either change the material or increase the thickness of the object to sustain the applied load.
- In this case, we change the material to **Alloy Steel**. To do so, right-click on the **Material** node in the **Analysis Manager**. A shortcut menu will be displayed as shown in Figure-23.

Figure-23. Shortcut menu for material

- Select the **Apply/Edit Material** from the menu. The **Material** dialog box will be displayed.
- Select the **Alloy Steel** material from the left table and click on the **Apply** button to apply the material.
- Click on the **Close** button from the dialog box to close it.
- Since the material is changed, we need to run the analysis again. To do so, click on the **Run** button from the **Ribbon**. The analysis will re-run and results will be displayed.
- Double-click on the **Factor of Safety1 (-FOS-)** node from the **Analysis Manager**. The results will be displayed as shown in Figure-24.

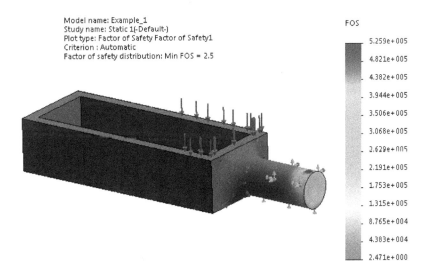

Figure-24. Updated Factor of Safety results

- Now, the factor of safety is **2.5** i.e. above **1**. So, our product can sustain the applied load.

SHELL MANAGER

You can manage all the shells/surfaces at a single place by using the **Shell Manager**. Shells are generally used to represent parts that have low thickness like sheet metal parts. The advantage of using shells lies in element type used for them during meshing. For solid objects 3D elements are used but for shells 2D elements like linear triangular and parabolic triangular elements are used. The analysis performed using 2D elements is faster. The procedure to use the **Shell Manager** is given next.

- Before using **Shell Manager**, you should have a surface model; refer to Figure-25. This model is a representation of sheet metal part with thickness of 3 mm at the top curved face and 5 mm on the other faces.

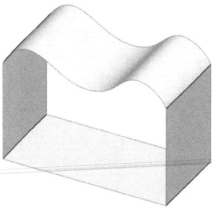

Figure-25. Surface model

- To assign thickness to the model, click on the **Shell Manager** tool from the **Simulation CommandManager** of the **Ribbon**. The **Shell Manager PropertyManager** will be displayed with **Shell Manager** pane; refer to Figure-26.

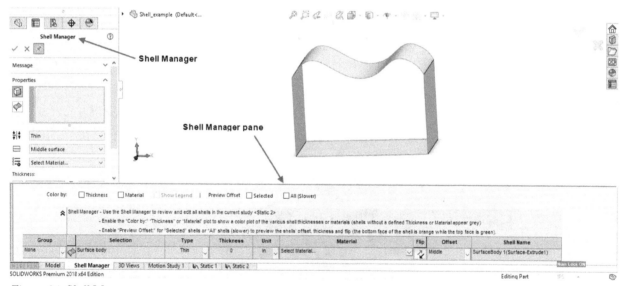

Figure-26. Shell Manager

- If the object has uniform thickness then click in the cell under **Thickness** column in the **Shell Manager** pane and enter the desired thickness value; refer to Figure-27. Make sure that you have selected correct unit from the **Unit** column in the pane.

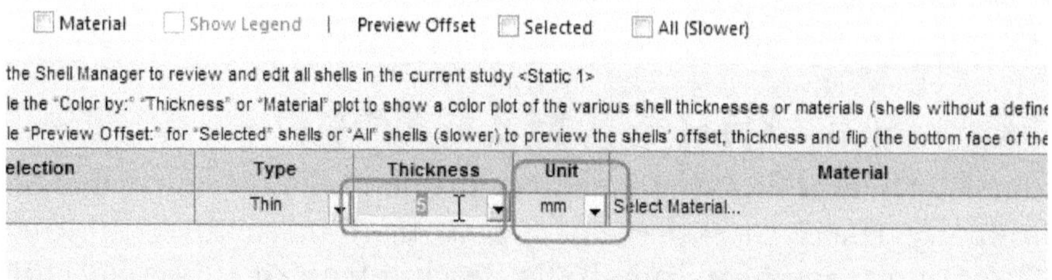

Figure-27. Uniform thickness applied

- If you want to specify different thickness to different faces then click in the selection box in **Properties** rollout of the **Shell Manager** and select the faces; refer to Figure-28.

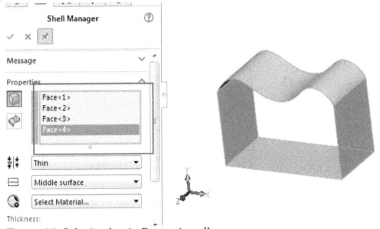

Figure-28. Selection box in Properties rollout

- Click on the **OK** button from the **Shell Manager**. A warning message will be displayed. Click on **Yes** to apply new selection of faces for shell. The faces will be displayed in the **Shell Manager** pane; refer to Figure-29.

Color by:	Thickness	Material	Show Legend		Preview Offset	Selected		All (Slower)		

Shell Manager - Use the Shell Manager to review and edit all shells in the current study <Static 1>
- Enable the "Color by:" "Thickness" or "Material" plot to show a color plot of the various shell thicknesses or materials (sh
- Enable "Preview Offset:" for "Selected" shells or "All" shells (slower) to preview the shells' offset, thickness and flip (th

Group		Selection	Type		Thickness	Unit		Material
None	▾	Face@Surface body	Thin	▾	0	mm	▾	Select Material...
None	▾	Face@Surface body	Thin	▾	0	mm	▾	Select Material...
None	▾	Face@Surface body	Thin	▾	0	mm	▾	Select Material...

Figure-29. Shell Manager pane

- One by one, enter the desired sheet thickness for each face. Make sure that you press **ENTER** after specifying thickness in the cells.
- Similarly, you can select other parameters like material, offset type and so on for individual faces. Figure-30 shows the faces assigned with different materials.

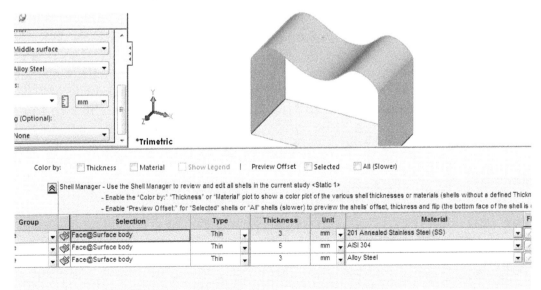

Color by:	Thickness	Material	Show Legend		Preview Offset	Selected		All (Slower)		

Shell Manager - Use the Shell Manager to review and edit all shells in the current study <Static 1>
- Enable the "Color by:" "Thickness" or "Material" plot to show a color plot of the various shell thicknesses or materials (shells without a defined Thickn
- Enable "Preview Offset:" for "Selected" shells or "All" shells (slower) to preview the shells' offset, thickness and flip (the bottom face of the shell is

Group		Selection	Type		Thickness	Unit		Material		Fl
:	▾	Face@Surface body	Thin	▾	3	mm	▾	201 Annealed Stainless Steel (SS)	▾	
:	▾	Face@Surface body	Thin	▾	5	mm	▾	AISI 304	▾	
:	▾	Face@Surface body	Thin	▾	3	mm	▾	Alloy Steel	▾	

Figure-30. Faces with different material and thickness

- After specifying the parameters, click on the **OK** button from the **Shell Manager**. Now, you can continue with the analysis just like a solid part (with hybrid material).

PERFORMING STATIC ANALYSIS USING STUDY ADVISOR

Earlier in this chapter, we have performed the static analysis by selecting each of the required option from the list of tools in the **Ribbon**. We can use the **Study Advisor** to perform all the steps related to analysis. The steps to use the Study Advisor are given next.

Starting the Static Analysis

- Open the part on which you want to perform the analysis.
- Click on the **SolidWorks Simulation** button from the **SOLIDWORKS Add-Ins** tab of the **Ribbon** to add **Simulation** tab in the **Ribbon**, if not added already.
- Click on the Down arrow below the **Study Advisor** button and select the **Study Advisor** tool from the drop-down. The study page of the **Simulation Advisor** will be displayed as shown in Figure-31.

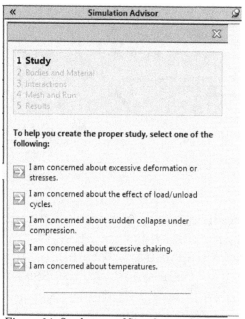

Figure-31. Study page of Simulation Advisor

- Currently, we are planning to perform the stress analysis. For that, click on the **I am concerned about excessive deformation or stresses** link button. The **second page of Study** will be displayed in the **Simulation Advisor**; refer to Figure-32.

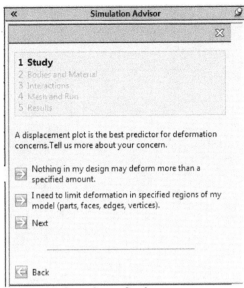

Figure-32. Second page Study

- Using the options in this page, we can check the deformation plot for the present analysis. Deformation plot displays the visualization of deformation for the analysis.

1. If you want to check the deformation for a specific part/face/edge/vertex then click on the **I need to limit deformation in specified regions of my model(parts, faces, edges, vertices)** link button. The **Sensor PropertyManager** will be displayed as shown in Figure-33.

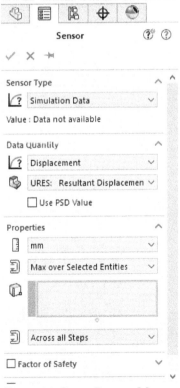

Figure-33. Sensor PropertyManager

2. Select the part/face/edge/vertex for which you want to check the deformation; refer to Figure-34 and click on the **OK** button from the **PropertyManager**. The page related to material failure will be displayed in the **Simulation Advisor**.

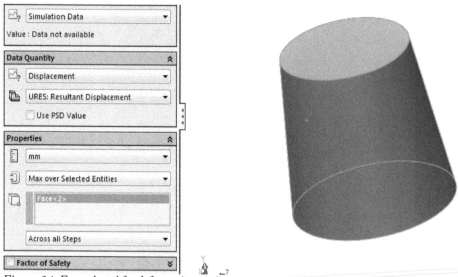

Figure-34. Face selected for deformation testing

- If you want to specify limit for deformation then click on the **Nothing in my design may deform more than a specified amount**. The **Sensor PropertyManager** will be displayed.

    1. Click on the **Alert** rollout to expand it. The options in the rollout will be displayed as shown in Figure-35.
    2. Click in the edit box and specify the limit of deformation. If the deformation crosses the specified value then an alert will be displayed.
    3. Click **OK** button from the **PropertyManager**. Page related to material failure will be displayed in the **Simulation Advisor**; refer to Figure-36.

Figure-35. Sensor PropertyManager with Alert options

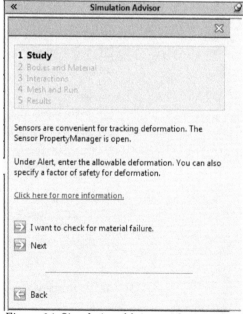

Figure-36. Simulation Advisor with page related to material failure

- Click on the **I want to check material failure** link button to check for material failure. The next page for material failure will be displayed in the **Simulation Advisor**; refer to Figure-37.

- Click on the **I don't know where to expect failure. Check the whole model** link button to check the whole model for material failure. You can select the **I am most concerned about failure at specific locations in my model** link button if you want to check the material failure for a specific portion. In this case, we have selected the first link button. On doing so, again the **Sensor PropertyManager** is displayed where you can specify the alert for maximum stress. Note that the maximum bearable stress is generally obtained by material properties.

- Click on the **OK** button from the **PropertyManager** and then click on the **Next** link button from the **Simulation Advisor**, the name of the study that suits your requirements will be displayed; refer to Figure-38.

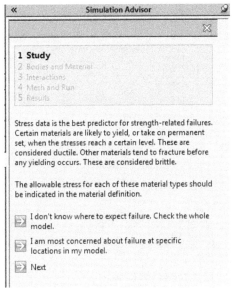

Figure-37. Simulation Advisor with another page related to material failure

Figure-38. Simulation Advisor with Study name

- Click on the **Create a Static Study** link button and then click on the **Next** link button from the **Simulation Advisor**. The **Bodies and Material** page will be displayed in the **Simulation Advisor**.

- Click on the **Next** button, the pages related to simulation support will be displayed; refer to Figure-40.

- Click on the **Next** button from the displayed **Simulation Advisor**. The page related to simulation symmetry will be displayed. If you apply symmetry then calculation time for analysis will be reduced.

- Click on the desired link from the **Simulation Advisor**. For the current example, click on the **Mirrored** link button and then click on the **Add symmetry fixture on shell edges** link button from the **Simulation Advisor**. The **Fixture PropertyManager** will be displayed; refer to Figure-41.

- Click on **Next** button from the displayed **Simulation Advisor** screen. The **Simulation** symmetry page will be displayed again. If you want to apply the symmetry then do as the procedure that is discussed earlier. Otherwise, click on the **Symmetry does not apply** button from **Simulation Advisor** page. The page related to body setup will be displayed; refer to Figure-42.

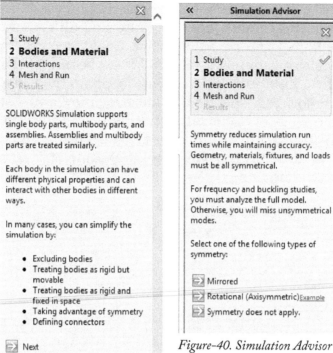

Figure-39. *Simulation Advisor with simulation support*

Figure-40. *Simulation Advisor with symmetry page*

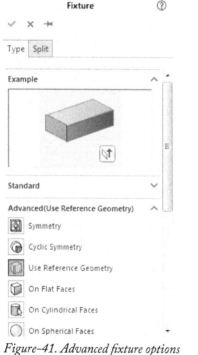

Figure-41. *Advanced fixture options*

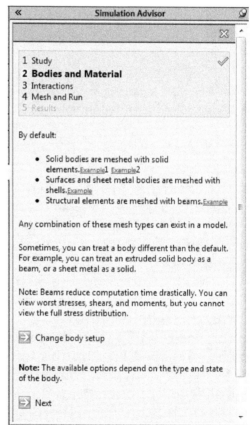

Figure-42. *Page related to body setup*

- Click on the **Change body setup** link button to specify setting for the body. You are displayed a message box saying that you need to right-click on the body and select the **Treat** as option. Right-click on the name of the model in the **Analysis**

Manager and select the desired treat as option; refer to Figure-43. The options related to the selected option will be displayed in the form of **PropertyManager**. Act accordingly.

- Now, click on the **Next** button from the body setup page of **Simulation Advisor**. The material page will be displayed; refer to Figure-44.

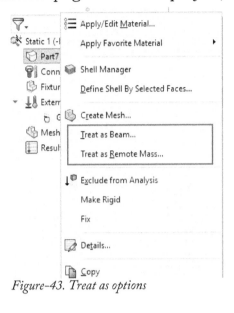

Figure-43. Treat as options

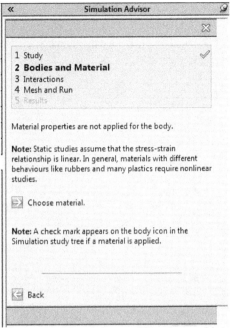

Figure-44. Material page of Simulation Advisor

- Click on the **Choose material** link button to specify the material for the current model. The **Material** dialog box will be displayed as discussed earlier. Select the material, click on the **Apply** button and then click on the **Close** button. The applied material will be displayed in the **Simulation Advisor**.
- Click on the **Next** button from the **Simulation Advisor**. The **Interactions** page will be displayed; refer to Figure-45.

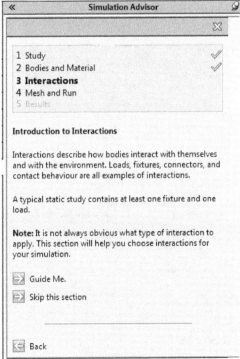

Figure-45. Simulation Advisor with Interaction page

From here onwards, the path of using **Simulation Advisor** becomes very confusing. So, we will go back to **Ribbon** and select the desired option from there.

- Click on the **Fixtures Advisor** button from the **Ribbon**. The options related to fixtures will be displayed in the **Simulation Advisor**; refer to Figure-46.
- Click on the **Add a fixture** link button. The **Fixture PropertyManager** will be displayed as discussed earlier.
- Select the bottom face of the current model; refer to Figure-47. And click on the **OK** button from the **PropertyManager** to apply the fixture.

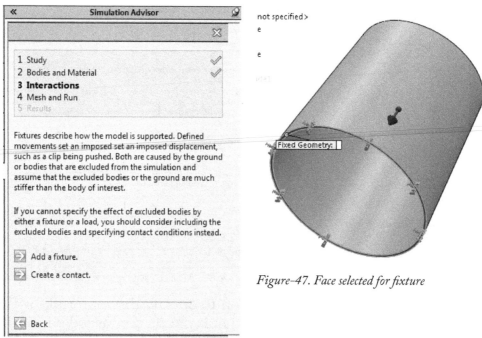

Figure-47. Face selected for fixture

Figure-46. Options for fixture in Simulation Advisor

- Click on the **External Loads Advisor** button from the **Ribbon**. The loads page of **Simulation Advisor** will be displayed; refer to Figure-48.

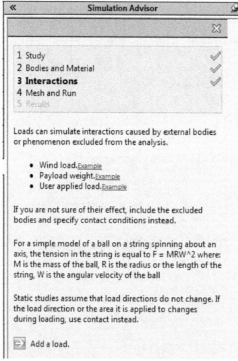

Figure-48. Loads page of Simulation Advisor

- Click on the **Add a load** link button. The **Force/Torque PropertyManager** will be displayed as discussed earlier.
- Select the top face of the model and specify the value of force as **392** N; refer to Figure-49.
- Click on the **OK** button from the **PropertyManager** to apply the force.
- Click on the **Done - Continue to next Advisor section** link button from the **Simulation Advisor**. The **Mesh and Run** page of **Simulation Advisor** will be displayed as shown in Figure-50.

Figure-49. Force applied on top face

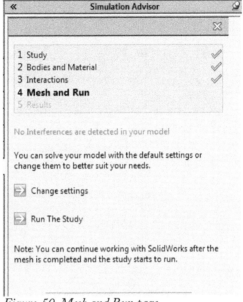

Figure-50. Mesh and Run page

- Click on the **Change settings** link button and then specify the desired mesh sizes for global or local areas of the model.
- In our case, we use the default settings and click on the **Run The Study** link button. First meshing will be processes and then the analysis will be performed. After the analysis is completed, then the results page will be displayed in the **Simulation Advisor** as shown in Figure-51.
- Click on the **Play** button from the **Simulation Advisor** to check the results. Click on the **Stop** button to stop animation.
- Click on the **Everything looks reasonable** link button to check the other results.
- Next, Click on the **Material breaking or yielding** link button and then **Show Factor of Safety plot** link button to plot the **Factor of Safety** result. The result will be displayed; refer to Figure-52.

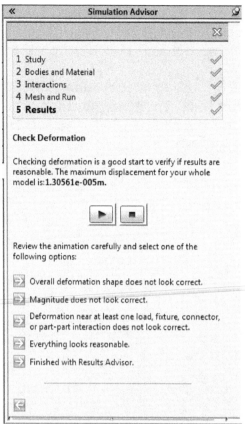

Figure-51. Result page of Simulation Advisor

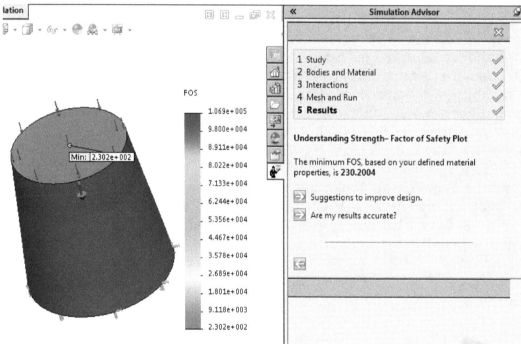

Figure-52. Factor of safety result

- Click on the **Suggestions to improve design** link button and you will get to know the methods to increase strength of the product if required; refer to Figure-53.

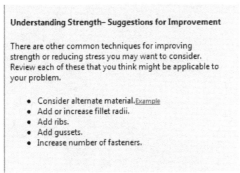

Figure-53. Suggestions to improve strength

- By default, the Von-mises stress plot is displayed but sometimes we need other plots of stress also. To do so, right-click on the **Results** node in the **Analysis Manager** and select the desired option like **Define Stress Plot** option. The **Stress Plot PropertyManager** will be displayed; refer to Figure-54.

Figure-54. Stress Plot Property-Manager

- Click on the **Component** drop-down and select the **P1: 1st Principal Stress** option from the list displayed.
- Click on the **OK** button from the **PropertyManager**. The plot will be created.

STATIC ANALYSIS ON ASSEMBLY

Problem:
We have an assembly of wall bracket as shown in Figure-55. The bracket is bolted to virtual wall with the help of foundation bolts. Load of 100 kg is applied at the flat end of the support. Check the component/components that are failing in the static analysis. The part files are available in the resources of this book.

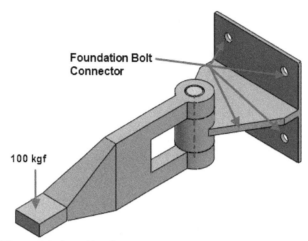

Figure-55. Assembly of wall bracket for static analysis

Understanding Problem:
In this case, we have an assembly of three components connected to each-other. The base plate is attached to a virtual wall with the help of foundation bolts. The arm is connected to base with the help of a pin. Now, we will apply these conditions in SolidWorks Simulation.

Starting Analysis

- Open the assembly file of model on which you want to perform the analysis.
- Click on the **SolidWorks Simulation** button from the **SOLIDWORKS Add-Ins** tab of the **Ribbon** to add **Simulation** tab in the **Ribbon**, if not added already.
- Click on the Down arrow below the **Study Advisor** button and select the **New Study** tool from the drop-down. List of analysis studies that can be performed, will be displayed.
- Click on the **Static** button if not selected. Specify the name of the analysis in the edit box displayed above the list of analyses.
- Click on the **OK** button from the **PropertyManager**. **Static 1** tab will be added at the bottom of the modeling area in the **StudyBar**. Also, the tools related to the analysis will be displayed in the **Ribbon**.

Applying Material to Parts of Assembly

There are two ways to apply material to the parts of assembly; Applying same material to all the components and applying different materials to components of the assembly. We will be applying different material to the assembly components.

- Right-click on the Base component from the **Parts** category in the **Analysis Manager**. A shortcut menu will be displayed; refer to Figure-56.

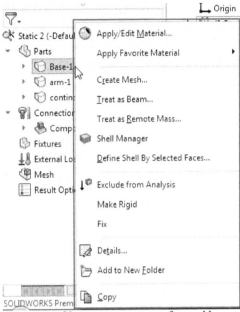

Figure-56. Shortcut menu parts of assembly

- Click on the **Apply/Edit Material** option from the shortcut menu. The **Material** dialog box will be displayed.
- Select the **Cast Alloy Steel** from the left area of the dialog box and click on the **Apply** button. Click on the **Close** button from the dialog box to exit.
- Similarly, apply alloy steel to arm and stud.

Note that if you want to apply same material to all the components in the assembly then right-click on the Parts category and click on the **Apply Material to All** option from the shortcut menu displayed; refer to Figure-57.

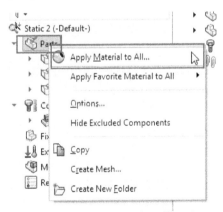

Figure-57. Apply Material to All option

Setting Global Contacts

By default, SolidWorks Simulation applies bonded contacts between all the components in the assembly which means all the components are perfectly bonded and behave as if they are single model entity. But, we need a no penetration contact in which all the components of assembly are separate and make affect on each other due to load applied. The procedure is discussed next.

- Expand the **Component Contacts** sub-category in the **Connections** category of the **Analysis Manager**; refer to Figure-58.
- Right-click on the **Global Contact** option and select the **Edit Definition** option from the shortcut menu displayed. The **Component Contact PropertyManager** will be displayed as shown in Figure-59.

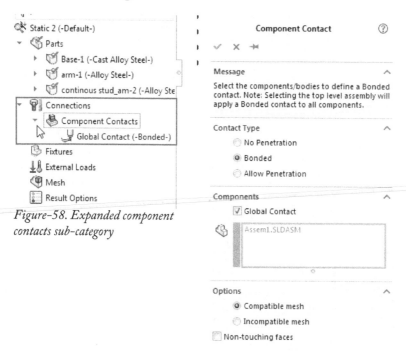

Figure-58. Expanded component contacts sub-category

Figure-59. Component Contact PropertyManager for assembly

- Select the **No Penetration** radio button from the **Contact Type** rollout in the **PropertyManager**.
- Click on the **OK** button from the **PropertyManager** to exit.

Setting Contact Manually

This section is just for information and does not affect the results of current simulation analysis. You can skip this section if you wish to. This tool may be important for other analysis but it is not for current analysis.

- Click on the **Contact Set** tool from the **Connections Advisor** drop-down in the **Ribbon**. The **Contact Sets PropertyManager** will be displayed; refer to Figure-60.
- Select the **Automatically find contact sets** radio button from the **PropertyManager** and select the **Touching faces** radio button to highlight all the contacts that are created by contacts of assembly parts.
- One by one select all the components of assembly for which you want to create the contact sets; refer to Figure-61.

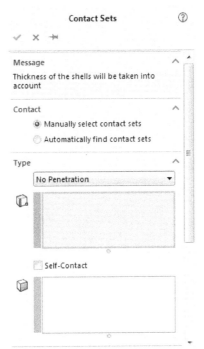

Figure-60. Contact Sets Property-Manager for assembly components

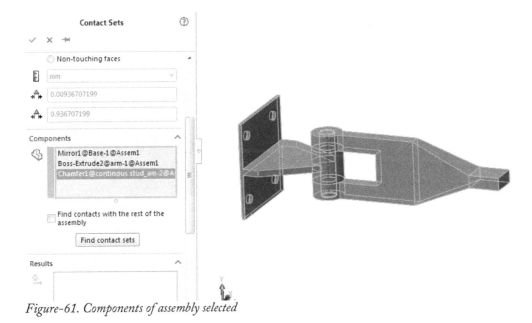

Figure-61. Components of assembly selected

- Click on the **Find contact sets** button in the **Components** rollout. The contact sets will be displayed in **Results** rollout.
- Select the contact set and set the desired condition in the **Type** drop-down.
- After setting all the contacts, click on the **OK** button from the **PropertyManager** to exit.

Creating Virtual Wall

As the name suggests, virtual wall acts as a virtual wall for the assembly. We will use it later to fasten foundation bolts.

- Click on the **Contact Sets** tool from the **Connections Advisor** drop-down in the **Ribbon**. The **Contact Sets PropertyManager** will be displayed.
- Select **Virtual Wall** option from the **Type** drop-down in the **Type** rollout of the **PropertyManager**; refer to Figure-62.

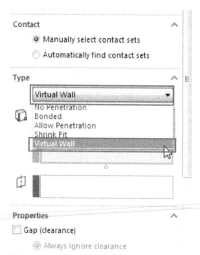

Figure-62. Virtual Wall option

- Select the flat back face of the base part and then click in the **Target Plane** selection box; refer to Figure-63.

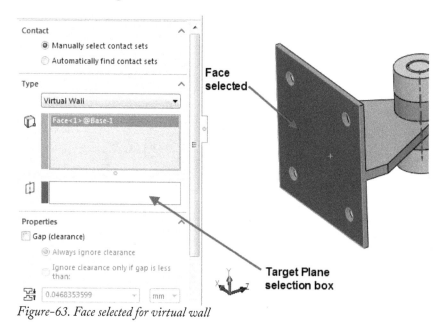

Figure-63. Face selected for virtual wall

- Expand the design tree from viewport and select the Front plane of Base part; refer to Figure-64.

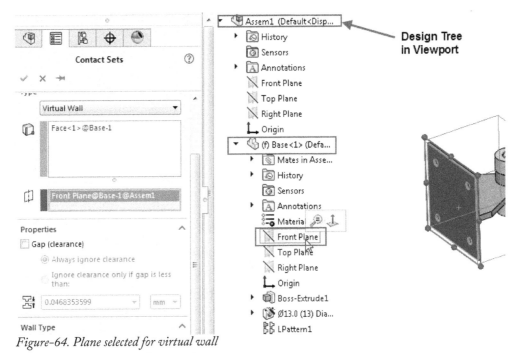

Figure-64. Plane selected for virtual wall

• Click on the **OK** button from the **PropertyManager**.

Creating Foundation Bolt connection

Foundation bolts give us the benefit of fixing any object to the wall/floor using bolts. The procedure to create connection is given next.

• Click on the **Bolt** tool from the **Connections Advisor** drop-down in the **Ribbon**. The **Connectors PropertyManager** will be displayed with the options related to bolts.
• Select the **Foundation Bolt** button from the **Type** rollout; refer to Figure-65.

Figure-65. Foundation Bolt button

• Select the round edge of a hole in the base plate; refer to Figure-66.

Edge selected

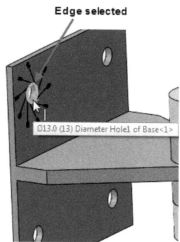

Ø13.0 (13) Diameter Hole1 of Base<1>

Figure-66. Edge selected for foundation bolt

- Click in the **Target Plane** selection box in the **PropertyManager** and select the **Front plane** of base part using the **Design Tree** in viewport.
- Move downward in **PropertyManager** and set the Torque value of pre-load 600 N.m.
- Click on the **OK** button from the **PropertyManager**. The foundation bolt will be created.
- Similarly, create the foundation bolts for other holes of base.

Applying Load

- Click on the **Force** tool from the **External Loads Advisor** drop-down in the **Ribbon**. The **Force/Torque PropertyManager** will be displayed.
- Select the flat face of the arm part and select the **Metric** option from the **Unit** drop-down in the **PropertyManager**; refer to Figure-67.

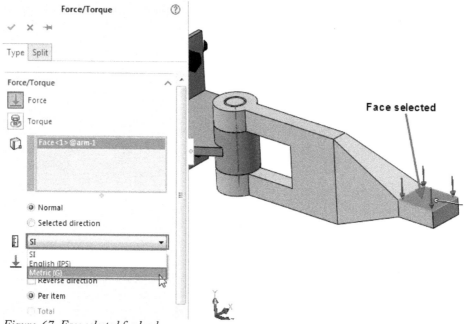

Figure-67. Face selected for load

- Specify the value of load as **100** in the **Force value** edit box in **PropertyManager**.
- Click on the **OK** button from the **PropertyManager**.

Meshing

- Click on the **Create Mesh** tool from the **Run** drop-down in the **Ribbon**. The **Mesh PropertyManager** will be displayed.
- Move the **Mesh Density** slider towards **Coarse**; refer to Figure-68 and click on the **OK** button to create meshing. This will reduce the processing time of analysis but reduce the accuracy also. If you have good system configuration and you are very serious about results of this analysis then move the slider towards **Fine**.

Figure-68. Mesh PropertyManager with Coarse density

Running Analysis

Before running analysis, we need to select the Direct Sparse solver for solving analysis because we have used the **No penetration contact** in our analysis. Procedure to select Direct sparse solver is given next.

- Right-click on the analysis name in **Analysis Manager**. A shortcut menu will be displayed; refer to Figure-69.

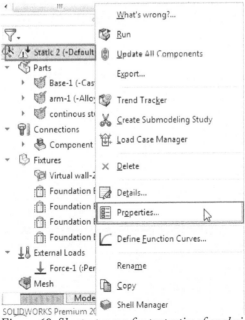

Figure-69. Shortcut menu for properties of analysis

- Click on the **Properties** option from the shortcut menu. The **Static** dialog box will be displayed with properties of analysis.
- Select the **Direct sparse solver** option from the drop-down in the **Solver** area of the dialog box; refer to Figure-70.

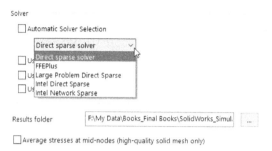

Figure-70. Direct sparse solver option

- Click on the **OK** button from the dialog box.

The use of different solvers is discussed next.

Automatic

The software selects the solver based on the study type, analysis options, contact conditions, etc. Select the **Automatic Solver Selection** check box to use this option. Some options and conditions apply only to either Direct Sparse or FFEPlus.

Direct Sparse

Select the Direct Sparse:
- when you have enough RAM and multiple CPUs on your machine.
- when solving models with No Penetration contact.
- when solving models of parts with widely different material properties.

For every 200,000 dof, you need 1 Gb of RAM for linear static analysis. The Direct Sparse solver requires 10 times more RAM than the FFEPlus solver.

FFEPlus (iterative)

The FFEPlus solver uses advanced matrix reordering techniques that makes it more efficient for large problems. In general, FFEPlus is faster in solving large problems and it becomes more efficient as the problem gets larger.

For every 2,000,000 dof, you need 1 Gb of RAM.

Large Problem Direct Sparse

By leveraging enhanced memory-allocation algorithms, the Large Problem Direct Sparse solver can handle simulation problems that exceed the physical memory of your computer.

If you initially select the Direct Sparse solver and due to limited memory resources it has reached an out-of-core solution, a warning message alerts you to switch to the Large Problem Direct Sparse.

The Large Problem Direct Sparse (LPDS) solver is more efficient than the FFEPlus and Direct Sparse solvers at taking advantage of multiple cores.

Intel Direct Sparse

The Intel Direct Sparse solver is available for static, thermal, frequency, linear dynamic, and nonlinear studies.

By leveraging enhanced memory-allocation algorithms and multi-core processing capability, the Intel Direct Sparse solver improves solution speeds for simulation problems that are solved in-core.

Intel Network Sparse

The **Intel Network Sparse** is used to solve the analysis on other network computers while keeping your system free for other operations. To use this solver, the host computer must have premium version of SolidWorks while the network computers should have SOLIDWORKS Simulation Worker Agent installed.

Select the **Average stresses at mid-nodes (high-quality solid mesh only)** check box if you want to average stresses at the mid points and then calculate the final stress value of model. This option improves the accuracy and solution time of analysis.

Now, we are ready to run the analysis.

* Click on the **Run** button from the **Ribbon** and check the results. You will get large displacement in the analysis. This is the stage where we need to change the desired model to sustain the load. When software tells you to use large displacement option then it means your model is undergoing plastic deformation.

One solution to this problem can be adding rib to the base part as shown in Figure-71. The changed base part is also available in the resource for the book. Replace it with the base part in assembly and re-run the analysis to see the magic.

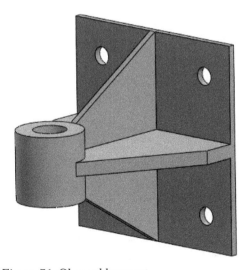

Figure-71. Changed base part

ADAPTIVE MESHING

There are three type of meshing available in SolidWorks Simulation; Standard Meshing, h-adaptive meshing, and p-adaptive meshing. Most of the time, the standard meshing does the work in static analyses. But, when we need more accurate results at deforming area then we use h-adaptive or p-adaptive meshing schemes. More detail on h-adaptive and p-adaptive is given next.

h-Adaptive Meshing

In h-adaptive meshing, system solves the analysis by using standard meshing and then system solves the analysis again with a refined mesh at the locations of strain. This process continues up to the number of steps defined by us or till the desired accuracy is achieved from the analysis. This repetition of analysis increases the processing time so this type of meshing is suggested when you are concerned about high accuracy. The procedure to apply h-adaptive meshing in the static study is given next.

*   After starting the **Static** study in SolidWorks Simulation, click on the **Study Properties** tool from the **Study Advisor** drop-down in the **Ribbon**; refer to Figure-72. The **Static** dialog box will be displayed as shown in Figure-73.

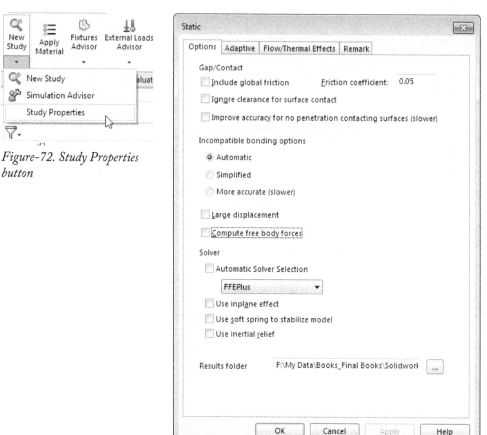

Figure-72. Study Properties button

Figure-73. Static dialog box

*   Click on the **Adaptive** tab from the dialog box. The options related to adaptive meshing will be displayed; refer to Figure-74.

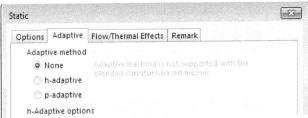

Figure-74. Partial view of adaptive options

- Click on the **h-adaptive** radio button from the **Adaptive method** area of the dialog box. The options related to h-adaptive meshing will become active; refer to Figure-75.

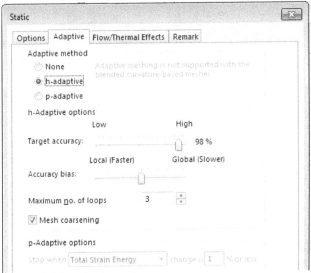

Figure-75. h-adaptive options

- Set the target accuracy by using the slider for **Target accuracy**.
- Set the accuracy bias using the **Accuracy bias** slider. If you want to concentrate on area under higher stress then move the slider towards left i.e. **Local(Faster)**. If you want to increase accuracy for the whole model then move the slider towards right i.e. **Global (Slower)**.
- Set the number of loops in the **Maximum no. of loops** spinner to set the number of times analysis should run for accuracy.
- Select the **Mesh coarsening** check box to make meshing coarse in the areas which are of no interest.
- Click on the **OK** button from the dialog box and run the analysis to check the difference in meshing and results. Figure-76 shows on such example.

Note that if the specified accuracy is not reached in number of loops specified in the dialog box then each time you click on the **Run** button to re-run the study, the results will become accurate further till the desired accuracy is achieved in case of h-adaptive meshing.

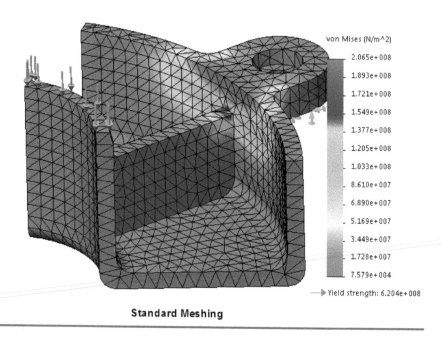

Standard Meshing

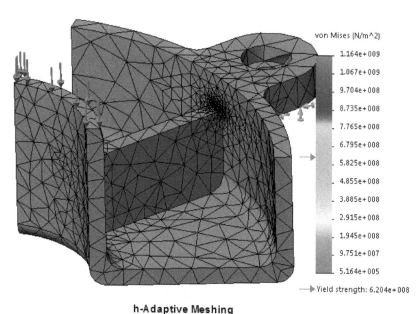

h-Adaptive Meshing

Figure-76. Standard and h-adaptive meshing

p-Adaptive Meshing

In h-adaptive meshing, we have seen that size of the mesh element reduces and number of elements increase in the area of interest. In p-Adaptive meshing, the order of mesh element changes. We have earlier discussed in this book that SolidWorks Simulation provides beam element for linear problems, triangular elements for 2D problems, and tetrahedral elements for 3D problems. So, in standard meshing highest order element is 2nd degree tetrahedral but this is not true if you have selected p-adaptive meshing. In this case, the elements can be from initial 2nd order to maximum 5th order depending on analysis requirements. The procedure to use the p-adaptive meshing is given next.

- Click on the **p-adaptive** radio button from the **Adaptive** tab in the **Static** dialog box discussed earlier. The options will be displayed as shown in Figure-77.

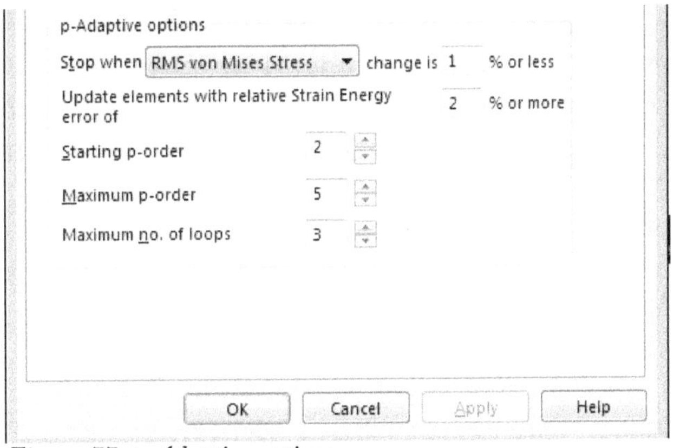

p-Adaptive options

Stop when | RMS von Mises Stress ▼ | change is 1   % or less

Update elements with relative Strain Energy
error of   2   % or more

Starting p-order   2

Maximum p-order   5

Maximum no. of loops   3

OK   Cancel   Apply   Help

Figure-77. p-Adaptive options

- Set the stopping condition for polynomial increment of mesh elements by using the drop-down and edit box in first line of p-Adaptive options area of the dialog box. There are three options in the drop-down; RMS Res. Displacement, RMS von Mises Stress, and Total Strain Energy. Select the desired option and set the percentage value of change to be achieved while making results accurate. If the change in selected parameters is less than this percentage then loop will not run further.

- The **Update elements with relative Strain Energy error of** edit box is used to specify the maximum allowed error value in the Strain energy results. The elements that are generating error more than this value will be upgraded.

- Specify the starting polynomial order and maximum polynomial order by using the respective spinners in the dialog box. Note that minimum polynomial order can be 2 and maximum polynomial order can be 5.

- Specify the number of loops up to which the analysis should run for accuracy. The maximum number of loops can be 4.

- After specifying the parameters, click on the OK button from the dialog box and run the analysis.

Figure-78 shows the example used for h-adaptive meshing solved by p-adaptive meshing.

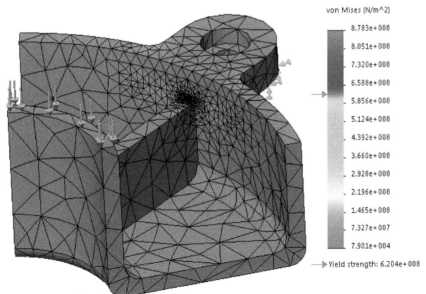

Figure-78. p-adaptive meshing results

SUBMODELING STUDY

We have worked earlier on an assembly of wall bracket and found that the reason of failure was base of wall bracket for which we added ribs as solution. But sometimes our components are hidden in the assembly and we are not able to check the stress without explicitly running analysis on them. To solve this problem, we use Submodeling study. In Submodeling study the selected portion of assembly is analyzed for the loads that are applicable on it during analysis of whole assembly. In this way, you do not require calculations of remote load. Note that this option is available for Static and Non-linear static analyses only.

Submodeling is based on the St. Venant's principle. The St. Venant's principle states that the stresses on a boundary reasonably distant from an applied load are not significantly altered if this load is changed to a statically equivalent load. The distribution of stress and strain is altered only near the regions of load application. You may cut a portion of the model, refine the mesh, and run analysis only for the selected portion provided that displacements are properly prescribed at the cut boundaries. If displacement results from the parent study are accurate, then these displacements are considered as boundary conditions at the cut boundaries for the submodeling study. So, may need to use adaptive meshing for accurate results of displacement. Note that the boundaries of the submodel must be adequately far from stress concentration areas.

There are a few conditions to use submodeling study like,

*   The study type must be static or nonlinear static with more than one body and not be a submodeling study itself. The parent study cannot be a 2D Simplification study.

*   The selected bodies of the submodel cannot have No penetration contact with unselected bodies that creates contact pressure across the cut boundary.
*   The selected bodies of the submodel cannot share connectors with unselected bodies. For example, the piping assembly below is not a suitable parent model. A submodel of two parts is connected to the excluded third part with bolts.
*   The cut boundary of the submodel cannot cut through bodies.
*   The cut boundary of the submodel cannot cut through a bonded contact defined by either beam-to-beam joints or shell edge-to-shell edge joints.
*   The bonded contact at the cut boundary of the submodel is formulated with an incompatible mesh.

The procedure to use submodeling study is discussed next through an example.

We have performed a static analysis on assembly given in Figure-79. Now, we want to check the results for Part 2 more precisely. To do so, we will create a submodeling study for which the steps are given next.

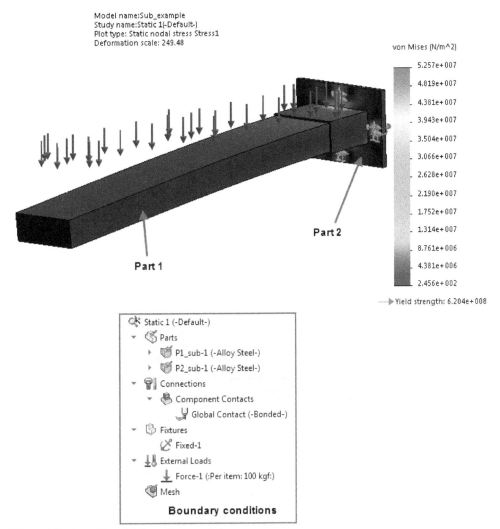

Model name:Sub_example
Study name:Static 1(-Default-)
Plot type: Static nodal stress Stress1
Deformation scale: 249.48

von Mises (N/m^2)

Part 2

Part 1

Yield strength: 6.204e+008

Boundary conditions

Figure-79. Assembly with boundary conditions

- Right-click on the analysis name in the **Analysis Manager** and click on the **Create Submodeling Study** option from the shortcut menu displayed; refer to Figure-80. The **Submodeling Information** dialog box will be displayed as shown in Figure-81.

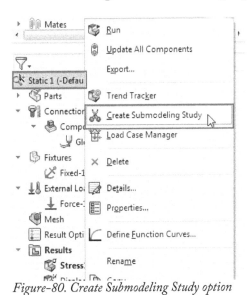

Figure-80. Create Submodeling Study option

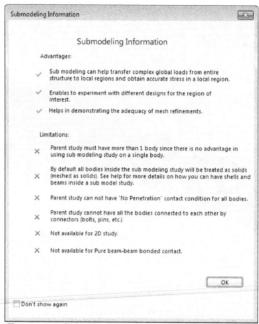

Figure-81. Submodeling Information dialog box

- Click on the **OK** button from the dialog box. You are asked to select the bodies that you want to include in submodeling study.
- Select the check box for the part that you want to include in study from the **Define Submodel PropertyManager**; refer to Figure-82.

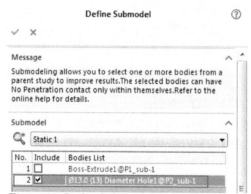

Figure-82. Define Submodel PropertyManager

- Click on the **OK** button from the **PropertyManager**. All the un-selected bodies will be excluded from the study and the study environment will be displayed; refer to Figure-83.

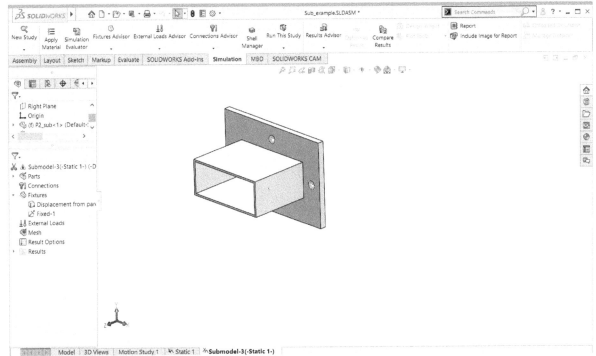

Figure-83. Submodeling study interface

- Click on the **Run This Study** button from the **Ribbon**. The result of assembly load will be displayed on the selected components; refer to Figure-84.

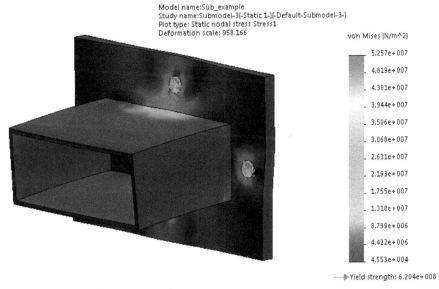

Figure-84. Result of submodel analysis

LOAD CASE MANAGER

Using **Load Case Manager**, we can create different cases of loading on the component and we can simultaneously judge the results of component in different loading conditions. This feature help us to find out the stability of component in entirely different conditions. For example, we need to test a component in a high temperature environment of 80 degree Celsius and a low temperature environment of -20 degree Celsius with different loads of 20 tonne and 15 tonne (I hope you know difference between ton and tonne). To do such analysis, we find load case manager very helpful. Note that load case manager is available for static analysis only and that too when p-adaptive and h-adaptive meshing is not selected. Now, we will discuss the use of load case manager with an example.

We have a static analysis performed on an alloy steel block; refer to Figure-85.

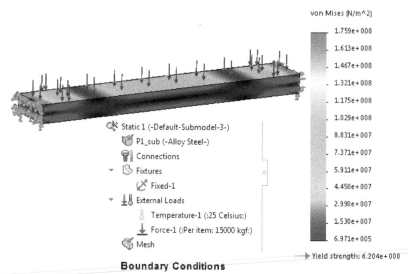

Figure-85. Static analysis with boundary conditions

- Right-click on the analysis name in the **Analysis Manager** and select the **Load Case Manager** option from the shortcut menu displayed; refer to Figure-86. The **Load Case Manager** will be displayed; refer to Figure-87.

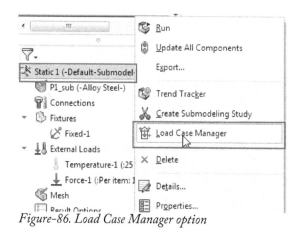

Figure-86. Load Case Manager option

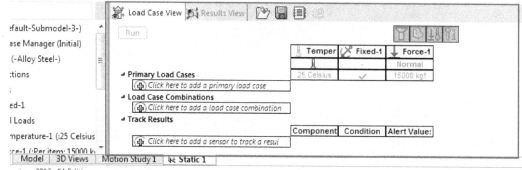

Figure-87. Load Case Manager

- Click on the toggle buttons in the Manager to display or hide fixtures, loads, and connectors.
- Click on the + sign under Primary Load Cases node. A new load case will be added; refer to Figure-88.

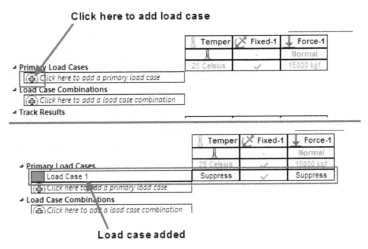

Figure-88. Adding load case

- Click on **Suppress** option in the field for which you want to specify new value. A drop-down will be displayed; refer to Figure-89.

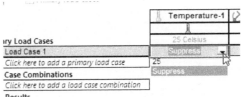

Figure-89. Suppress option for loads

- Select the earlier specified value from the drop-down and specify the new value like 80.
- Similarly, specify value of other loads in the respective fields.
- After specifying the values of case 1, again click on the **+** sign in the Primary Load Case node and repeat the procedure to specify loads for second case.
- You can combine various load cases by using the **Load Case Combinations** node options. Click on the **+** sign under **Load Case Combinations** node. The **Edit Equation (Load Case Combination)** dialog box will be displayed; refer to Figure-90.

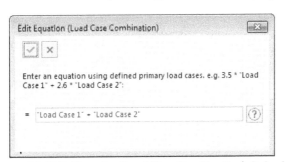

Figure-90. Edit Equation(Load Case Combination) dialog box

- Set the desired loading equation and click on the **OK** button to combine various loading cases.
- After setting the desired parameters, click on the **Run** button from the **Load Case Manager**. The system will solve various cases and display results in viewport.
- To check the result of a loading case, click on its name from **Results View** tab in the **Load Case Manager**; refer to Figure-91.

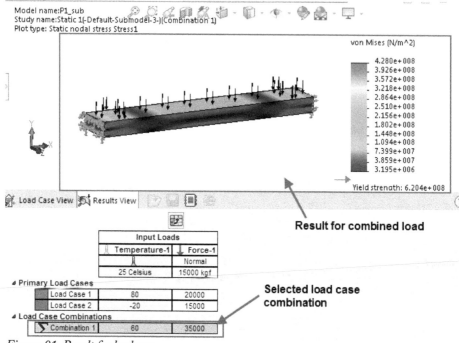

Figure-91. Result for load case

- Click on the **Report** button to create report for loading cases.
- Click on the **Close** button at the top-right of **Load Case Manager** to close it.

TREND TRACKER

The **Trend Tracker** tool is used to keep track of changes in major properties of the component under analysis. The tool is available for Static Analysis in shortcut menu of analysis; refer to Figure-92. The procedure to use this tool is given next.

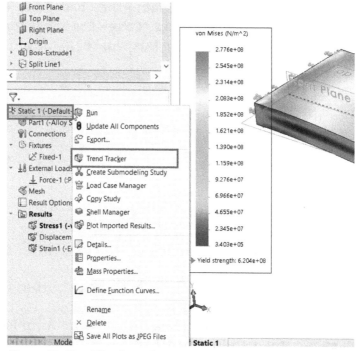

Figure-92. Trend Tracker tool

- Right-click on the **Static** option from the **Analysis Tree** after performing the analysis. The shortcut menu will be displayed; refer to Figure-92.
- Click on the **Trend Tracker** tool from the shortcut menu. The Trend Tracker option will be added in the **Analysis Tree**; refer to Figure-93.

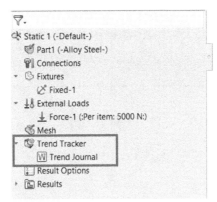

Figure-93. Trend Tracker node added in Analysis Tree

- Right-click on **Trend Tracker** node in the **Analysis Tree** and select the **Set Baseline** option. The current results of the analysis will be set as baseline for graphs recorded in the word document Trend Journal.
- Right-click on **Trend Tracker** node in the **Analysis Tree** again and select the **Manually add iteration** option if you want to add iterations manually otherwise they will be added automatically in the journal.
- Now, modify the parameters of model based on your requirement like reducing/ increasing size and perform the static analysis. The options in **Trend Tracker** node will be displayed as shown in Figure-94.

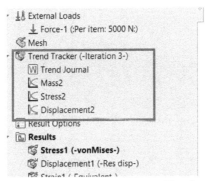

Figure-94. Modified options in Trend Tracker node

- Double-click on the **Trend Journal** option from the **Trend Tracker** node to check the trend of results in a word document.

Note that there might be some situations when you have sharp edges in the model that cause singularity then you should go for convergence in displacement in the results. Because, the stress will keep on going higher and higher while you keep on reducing the size of mesh element. When the element size will tend to 0, the stress in singularity area will tend to go infinity.

PRACTICE 1

Consider a rectangular plate with cutout. The dimensions and the boundary conditions of the plate are shown in Figure-95. It is fixed on one end and loaded on the other end. Under the given loading and constraints, plot the deformed shape. Also, determine the principal stresses and the von Mises stresses in the bracket. Thickness of the plate is 0.125 inch and material is **AISI 1020**.

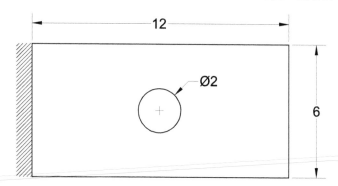

Figure-95. Drawing for Practice 1

PRACTICE 2

Download the model for practice 2 of chapter 3 from the resource kit and perform the static analysis using the conditions given in Figure-96. Find out the **Factor of Safety** for the model.

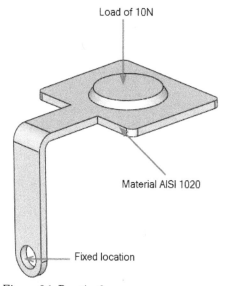

Figure-96. Practice 2

PRACTICE 3

Download the model of chair in chapter 3 from the resource kit and perform the static analysis using the conditions given in Figure-97. The material used for manufacturing chair is PET (Polyethylene terephthalate). Check whether it is safe. Also, check whether Rikishi (wrestler) can sit on this chair without any consequences. His weight is 193 Kg.

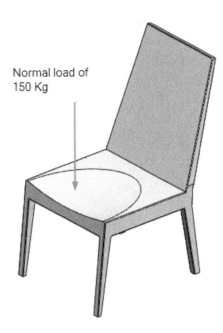

Figure-97. Practice 3 for Static Analysis

SELF-ASSESSMENT

Q1. Which of the following is not an assumption for static analysis?

a. The loads applied do not vary with time.
b. The change in stiffness due to loading is neglected.
c. Material properties are not affected by load rate.
d. Weld material and the heat affected zone will be neglected.

Q2. Which of the following is not an assumption for static analysis?

a. Residual stress due to fabrication, pre-loading of bolts, welding and/or other manufacturing or assembly processes are neglected.
b. Bolt loading is primarily axial in nature.
c. Failure of fasteners will be reflected in the analysis.
d. Surface torque loading due to friction will produce only local effects.

Answer the following questions in context of general analysis assumptions:

Q3. If the results in the particular area are of interest, then mesh convergence should be limited to this area. (T/F)

Q4. No slippage between interfacing components will be assumed. (T/F)

Q5. Stiffness of bearings in radial or axial directions will be considered infinite. (T/F)

Q6. When a system under load is analyzed with linear static analysis, the linear finite element equilibrium equations are solved to calculate the displacement components at all nodes. (T/F)

Q7. For solid objects 3D elements are used but for shells 2D elements like linear triangular and parabolic triangular elements are used. (T/F)

Q8. You can manage all the shells at a single place by using the .... tool from **Ribbon**.

Q9. By default, SolidWorks Simulation applies .... contacts between all the components in the assembly.

Q10. .... bolts give us the benefit of fixing any object to the wall/floor using bolts.

FOR STUDENT NOTES

Answers to Self-Assessment :
1. d, 2. c, 3. T, 4. T, 5. T, 6. T, 7. T, 8. Shell Manager, 9. bonded, 10. Foundation

Chapter 5

Non-Linear
Static Analysis

Topics Covered

The major topics covered in this chapter are:

- *Introduction*
- *Starting Non-Linear Static Analysis.*
- *Applying Material.*
- *Defining Fixtures*
- *Applying loads*
- *Defining Connections*
- *Simulating Analysis.*
- *Interpreting results*

NON LINEAR STATIC ANALYSIS

All real structures behave nonlinearly in one way or another at some level of loading. In some cases, linear analysis may be adequate. In many other cases, the linear solution can produce erroneous results because the assumptions upon which it is based are violated. Nonlinearity can be caused by the material behavior, large displacements, and contact conditions.

You can use a nonlinear study to solve a linear problem. The results can be slightly different due to different assumption.

In nonlinear finite element analysis, a major source of nonlinearity is due to the effect of large displacements on the overall geometric configuration of structures. Structures undergoing large displacements can have significant changes in their geometry due to load-induced deformations which can cause the structure to respond nonlinearly in a stiffening and/or a softening manner.

For example, cable-like structures generally display a stiffening behavior on increasing the applied loads while arches may first experience softening followed by stiffening, a behavior widely-known as the snap-through buckling.

Another important source of nonlinearity stems from the nonlinear relationship between the stress and strain which has been recognized in several structural behaviors. Several factors can cause the material behavior to be nonlinear. The dependency of the material stress-strain relation on the load history (as in plasticity problems), load duration (as in creep analysis), and temperature (as in thermoplasticity) are some of these factors.

This class of nonlinearity, known as material nonlinearity, can be idealized to simulate such effects which are pertinent to different applications through the use of constitutive relations.

A special class of nonlinear problems is concerned with the changing nature of the boundary conditions of the structures involved in the analysis during motion. This situation is encountered in the analysis of contact problems.

Pounding of structures, gear-tooth contacts, fitting problems, threaded connections, and impact bodies are several examples requiring the evaluation of the contact boundaries. The evaluation of contact boundaries (nodes, lines, or surfaces) can be achieved by using gap (contact) elements between nodes on the adjacent boundaries.

Yielding of beam-column connections during earthquakes is one of the applications in which material nonlinearity are plausible.

THEORY BEHIND NON-LINEAR ANALYSIS

As you have learnt from introduction, there are three factors which cause non-linearity in the body which are elastic behavior of material, large displacement in body, and variation in contact. In terms of equation this can be expressed as

$$[K(D)]\{D\} = \{F\}$$

Here, K(D) is stiffness matrix which is function of displacement.

D is the displacement matrix
and   F is the Force applied

So, from the above equation you can easily understand that the stiffness is changing according to the displacement. Note that displacement is vector and it also has a direction. If we compare the linear static analysis and non-linear static analysis curves then we can find out the different effects of load in both conditions; refer to Figure-1.

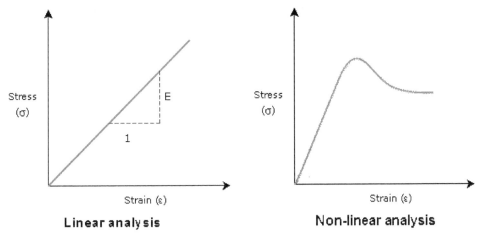

Figure-1. Stress Strain diagrams

In terms of load and displacement, the curve for both analysis can be given by Figure-2.

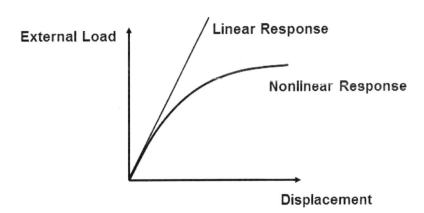

Figure-2. Load displacement curve

The normal incremental iteration does not work good for the non-linear analysis and generate errors as shown in Figure-3. So, Newton-Raphson algorithm is used to solve the non-linear equation.

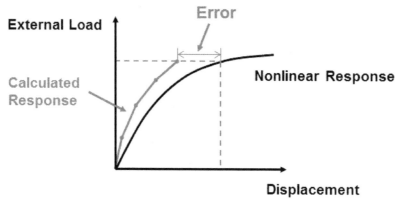

Figure-3. Error in incremental interative method

The equation for Newton-Raphson method is given as:

$$[K_T]\{\Delta u\} = \{F\} - \{F^{nr}\}$$

$[K_T]$ = tangent stiffness matrix
$\{\Delta u\}$ = Displacement increment
$\{F\}$ = external load vector
$\{F^{nr}\}$ = internal force vector

The iteration continues till $\{F\}$ - $\{F^{nr}\}$ (difference between external and internal loads) is within a tolerance; refer to Figure-4.

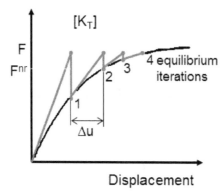

Figure-4. Newton-Raphson method

Thus a nonlinear solution typically involves the following:

- One or more load steps to apply the external loads and boundary conditions.(This is true for linear analyses too.)
- Multiple sub-steps to apply the load gradually. Each sub-step represents one load increment.(A linear analysis needs just one sub-step per load step.)
- Equilibrium iterations to obtain equilibrium (or convergence) at each sub-step. (Does not apply to linear analyses.)

Role of Time in Non-Linear Analysis

Each load step and sub-step is associated with a value of time. Time in most nonlinear static analyses is simply used as a counter and does not mean actual, chronological time. Figure-5 shows a load-time curve.

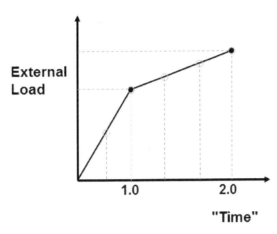

Figure-5. Load time curve

By default, time = 1.0 at the end of load step 1, 2.0 at the end of load step 2, and so on. For rate-independent analyses, you can set it to any desired value for convenience. For example, by setting time equal to the load magnitude, you can easily plot the load-deflection curve.

The "time increment" between each sub-step is the time step Δt. Time step Δt determines the load increment ΔF over a sub-step. The higher the value of Δt, the larger the ΔF, so Δt has a direct effect on the accuracy of the solution. SolidWorks Simulation gives the flexibility to use automatic time stepping algorithm and manual setting of time steps.

Now, we are going to start with Non-linear Static Analysis which means there is no damping or resistance to the force. Note that the last example in previous chapter, an assembly of wall bracket, was a problem of Nonlinear static analysis because there was large displacement in the body. Now, we will learn the procedures of SolidWorks Simulation to perform Nonlinear static analysis.

Starting Nonlinear Static Analysis

- Open the part on which you want to perform the analysis.
- Click on the **SolidWorks Simulation** button from the **SOLIDWORKS Add-Ins** tab of the **Ribbon** to add **Simulation** tab in the **Ribbon**, if not added already.
- Click on the Down arrow below the **Study Advisor** button and select the **New Study** tool from the drop-down. List of analysis studies that can be performed, will be displayed in the left.
- Click on the **Nonlinear** button and then on the **Static** button; refer to Figure-6.

Figure-6. Study PropertyManager

- Click on the **OK** button from the **PropertyManager**. The tools related to **Non-linear Static** analysis will be displayed. Note that most of the tools are same as discussed in previous chapters.

Applying Material

- Click on the **Apply Material** button from the **Ribbon**. The **Material** dialog box will be displayed as discussed in earlier chapters.
- Select the **AISI 1035 Steel(SS)** option from the left list and click on the **Apply** button.
- Click on the **Close** button from the dialog box to exit. The material will be applied; refer to Figure-7.

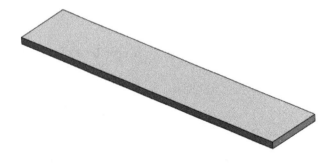

Figure-7. Model after applying material

Applying Fixtures

- Click on the down arrow below **Fixtures Advisor** in the **Ribbon**. The tools related to fixtures will be displayed.
- Click on the **Fixed Geometry** button from the tool list. The **Fixture PropertyManager** will be displayed; refer to Figure-8.

Figure-8. Fixture PropertyManager

- Select the left flat faces of the model; refer to Figure-9.

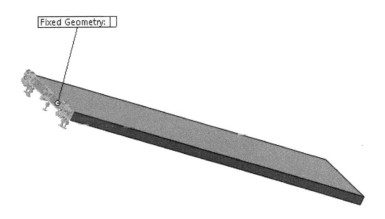

Figure-9. Faces fixed for analysis

- Click on the **OK** button from the **PropertyManager**. The holes will be fixed as if they are bolted at same position.

Applying Time Varying Force

- Click on the down arrow below **External Loads Advisor** button in the **Ribbon**. List of the tools related to loads will be displayed.
- Click on the **Force** button from the list. The **Force/Torque PropertyManager** will be displayed; refer to Figure-10.

- Note that **Variation with Time** rollout is added in the **PropertyManager** and two radio buttons are displayed named, **Linear** and **Curve**. On selecting the **Linear** radio button, the force will gradually increase till it reaches to the specified force value within the **1** second of time span; refer to Figure-11.

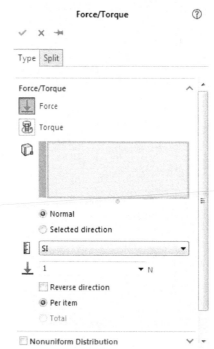

Figure-10. Force Torque PropertyManager

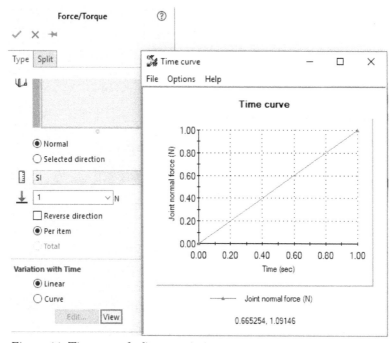

Figure-11. Time curve for linear variation

- Click on the **Curve** radio button from the **Variation with Time** rollout. The **Edit** button will become active.
- Click on the **Edit** button. The **Time curve** dialog box will be displayed as shown in Figure-12.

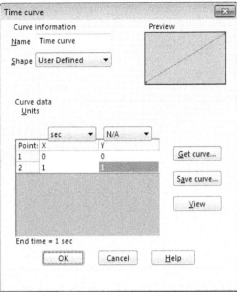

Figure-12. Time curve dialog box

- Click in the 2nd row of first column and specify the time as 20 in the field. In the same way, click in the 2nd row of the 2nd column and specify the force value as 10; refer to Figure-13.

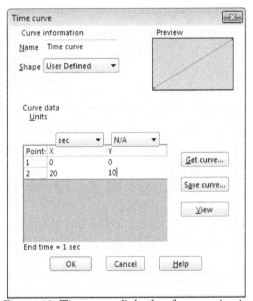

Figure-13. Time curve dialog box for customization

- Click on the **OK** button from the dialog box and then select the edge as shown in Figure-14 to apply the force.

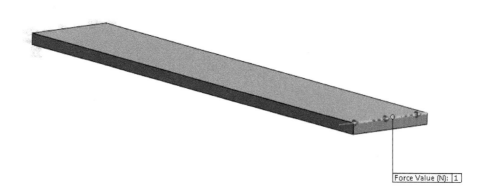

Figure-14. Edge selected for applying force

- Click on the **Selected direction** radio button from the **Force/Torque** rollout. You are asked to select a direction reference.
- Select the top face of the model; refer to Figure-15.

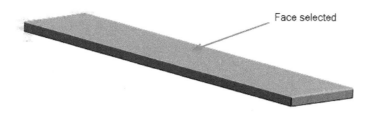

Figure-15. Face selected for force direction

- Now, click on the third button i.e. **Normal to Plane** and specify the force value as **10** N; refer to Figure-16.
- Note that if you click on the **View** button from the **Variation with Time** rollout; the graph will be displayed with the total force of 10 x 10= **100** N; refer to Figure-17.

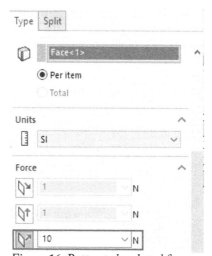

Figure-16. Button to be selected from Force rollout

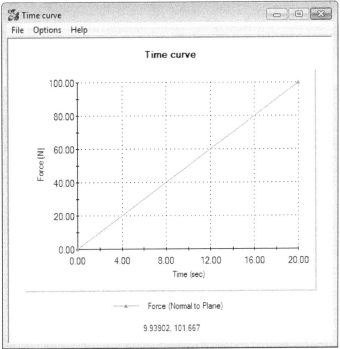

Figure-17. Time curve dialog box after entering 10 in force edit box

- Close the dialog box and click on the **OK** button from the **PropertyManager**. The force will be applied.

Running the analysis

- Click on the **Run** button from the **Ribbon**. The system will start analyzing the problem. Once the system is done, the results will be displayed in the Modeling area; refer to Figure-18. Note that from SolidWorks 2015 onwards, you can also see the intermediate results by selecting the **Pause** button from the **Solver** dialog box. Make sure that you have selected the **Show intermediate results upto current step** check box to display the results; refer to Figure-19. If you find that results are not as per expectation then click on the **Cancel** button and perform the changes.

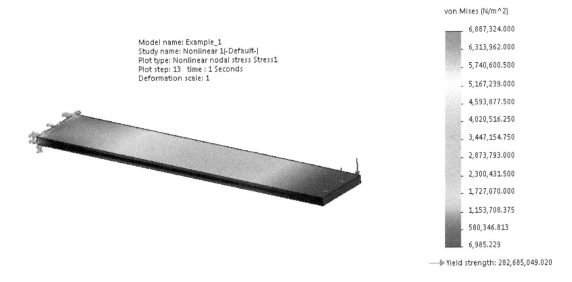

Figure-18. Result

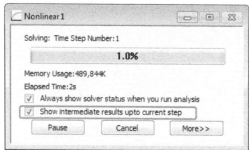

Figure-19. Non linear analysis solver dialog box

You can interpret the results as discussed in previous chapter.

SETTING PROPERTIES FOR NON-LINEAR STATIC ANALYSIS

It is very important to understand the basic properties of Non-linear analysis and how the changes in properties affect the analysis. The steps below will take you through various property settings.

- After starting Non-linear static analysis, click on the **Study Properties** tool from the **Study Advisor** drop-down in the **Ribbon**. The **Nonlinear-Static** dialog box will be displayed as shown in Figure-20.

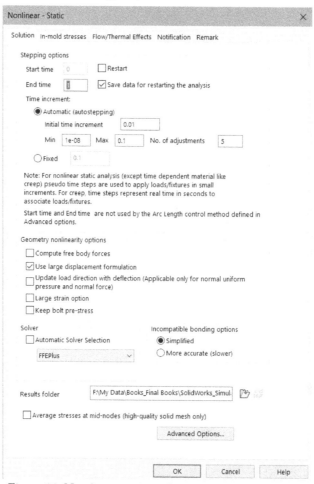

Figure-20. Non-linear static analysis properties dialog box

The options in **Stepping options** area of the dialog box are used to modify the start time, end time, and incremental steps. For example, you can check the effect of load on a component starting at 0 second of loading till 15 second after loading. Also, you can check the effect of loading after every 3 seconds.

- Note that the start time of analysis is fixed. To change the end time for analysis, click in the **End time** edit box and change the value as required (say 15 second).
- To change the value of time increment, select the **Fixed** radio button and specify the desired value in the edit box.
- The options in the **Geometric nonlinearity options** area are used to modify options related to geometric nonlinearity. There are three check boxes in this area. Select the **Use large displacement formulation** check box to use the formula of large displacement. It is important to select this check box when you know that the stiffness matrix will change according to the loading. Select the **Update load direction with deflection (Applicable only for normal uniform pressure and normal force)** check box if you want to keep pressure/force applied on the always normal to selected face even when there is large deformation in the model. Select the **Large strain option** check box if you are checking the deflection of plastic materials. Select the **Keep bolt pre-stress** check box to include prestressing of bolt during analysis process.
- Select the desired solver as discussed earlier.
- If you want to use pre-stresses of SolidWorks Plastics then click on the **In-mold stresses** tab in the dialog box and select the **Import in-mold stresses from SOLIDWORKS Plastics** check box to activate the option. Click on the **Browse** button and select the exported file of SolidWorks Plastics.
- Select the **Include material from SOLIDWORKS Plastics** check box to include the material of component in current analysis from SolidWorks Plastics.
- Click on the **OK** button from the dialog box to apply settings.

PERFORMING NON-LINEAR STATIC ANALYSIS ON AN ASSEMBLY

For performing non-linear static analysis on an assembly, we need to apply the contacts between various components in the assembly. The procedure to perform non-linear analysis on an assembly is given next.

Starting Non-Linear Static Analysis

- Open the part in SolidWorks on which you want to perform the analysis.
- Click on the **SolidWorks Simulation** button from the **SOLIDWORKS Add-Ins** tab of the **Ribbon** to add **Simulation** tab in the **Ribbon**, if not added already.
- Click on the Down arrow below the **Study Advisor** button and select the **New Study** tool from the drop-down. List of analysis studies that can be performed, will be displayed in the left.
- Click on the **Nonlinear** button and then on the **Static** button from the **Options** rollout.
- Click on the **OK** button from the **PropertyManager**. The tools related to **Non-linear Static** analysis will be displayed.

Applying the Material

- Click on the **Apply Material** button from the **Ribbon**. The **Material** dialog box will be displayed.

- Select the **Alloy Steel** material from the left and click on the **Apply** button. Note that the elastic properties of material will play an important role in the analysis.
- Click on the **Close** button from the dialog box. The material will be applied.

Applying Fixtures

- Click on the down arrow below **Fixtures Advisor** in the **Ribbon**. The tools related to fixtures will be displayed.
- Click on the **Fixed Geometry** button from the tool list. The **Fixture PropertyManager** will be displayed.
- Select the faces as shown in Figure-21.

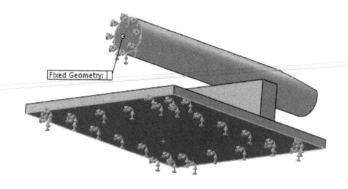

Figure-21. Faces selected for fixing

- Click on the **OK** button from the **PropertyManager** to fix the faces.

Applying Forces

- Click on the down arrow below **External Loads Advisor** button in the **Ribbon**. List of the tools related to loads will be displayed.
- Click on the **Force** button from the list. The **Force/Torque PropertyManager** will be displayed.
- Note that **Variation with Time** rollout is added in the **PropertyManager** and two radio buttons are displayed named, **Linear** and **Curve**. Make sure that the **Linear** radio button is selected in the rollout.
- Enter the value of force as **10000** N in the **Force Value** edit box.
- Select the face as shown in Figure-22.
- Click on the **OK** button from the **PropertyManager** to apply the force.

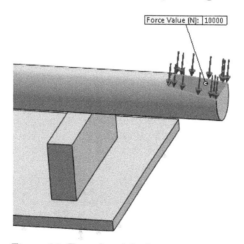

Figure-22. Face selected for force

Applying Connections

- Expand the **Component Contacts** node and right-click on the **Global Contact (-Bonded-)** option; refer to Figure-23.

Figure-23. Option to be selected

- Select the **Delete** option from the shortcut menu. The **Simulation** dialog box will be displayed.
- Click on the **Yes** button to delete the contact.
- Click on the down arrow below the **Connections Advisor** button and select the **Component Contact** button; refer to Figure-24. The **Component Contact PropertyManager** will be displayed as shown in Figure-25.

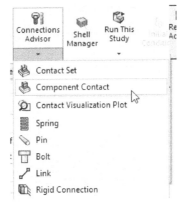

Figure-24. Component Contact button

Component Contact   ⑦

Message ∧

Select the components/bodies to define a Bonded contact. Note: Selecting the top level assembly will apply a Bonded contact to all components.

Contact Type ∧
- ⦿ Bonded
- ⦾ Allow Penetration

Components ∧
- ☐ Global Contact

Options ∧
- ⦿ Compatible mesh
- ⦾ Incompatible mesh
- ☐ Non-touching faces

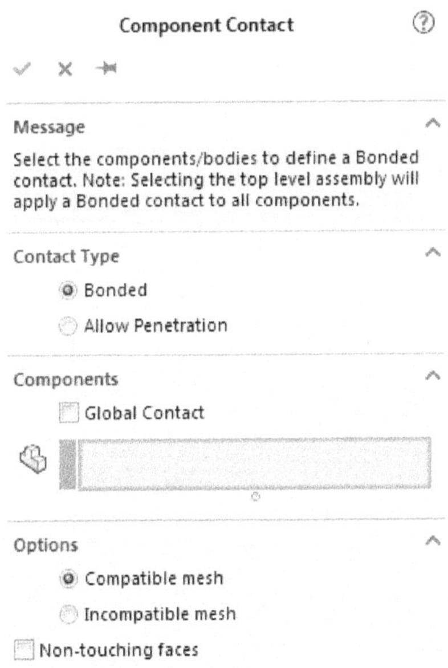

Figure-25. Component Contact Property Manager

- Select the two components of the assembly as shown in Figure-26.

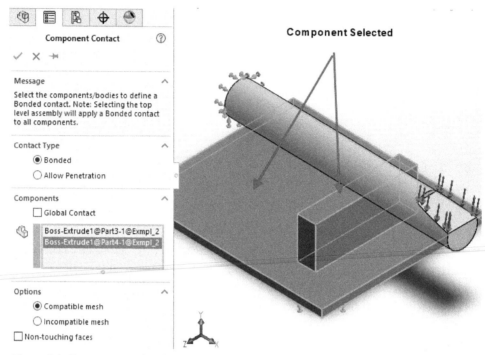

Figure-26. Components to be selected

- Click on the **OK** button from the **PropertyManager**. The contact will be added between the two components.
- Again, click on the down arrow below the **Connections Advisor** button and click on the **Contact Set** button. The **Contact Sets PropertyManager** will be displayed as shown in Figure-27.

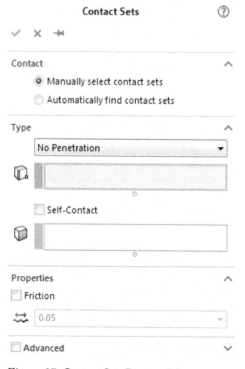

Figure-27. Contact Sets PropertyManager

- Select the round face of the first part in assembly.
- Click in the pink box in the **Type** rollout and select the flat faces of the block; refer to Figure-28.

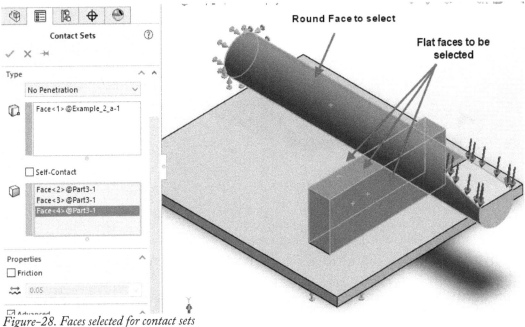

Figure-28. Faces selected for contact sets

- Select the check box next to **Advanced** rollout. The rollout will expand.
- Select the **Surface to surface** radio button.
- Click on the **OK** button from the **PropertyManager**. The contacts will be applied.

Running the analysis

- Click on the **Run** button from the **Ribbon**. The analysis will start.
- Note that since the faces that we selected later are not in contact to each other, so the dialog box with the related message; refer to Figure-29.

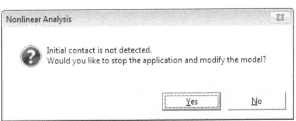

Figure-29. Nonlinear Analysis dialog box

- Click on the **No** button from the dialog box since we want to run the analysis with these conditions only. System will start solving the analysis.
- On completion of the analysis, the result will be displayed; refer to Figure-30.

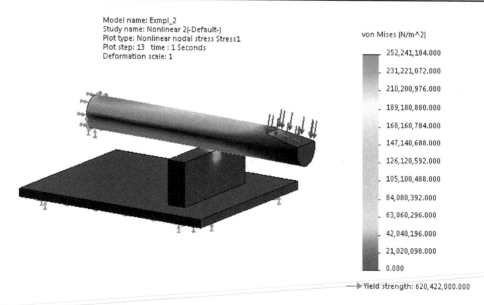

Model name: Exmpl_2
Study name: Nonlinear 2(-Default-)
Plot type: Nonlinear nodal stress Stress1
Plot step: 13   time : 1 Seconds
Deformation scale: 1

von Mises (N/m^2)

252,241,184.000
231,221,072.000
210,200,976.000
189,180,880.000
168,160,784.000
147,140,688.000
126,120,592.000
105,100,488.000
84,080,392.000
63,060,296.000
42,040,196.000
21,020,098.000
0.000

Yield strength: 620,422,000.000

Figure-30. Analysis result

- Note that the stress is also induced in the block below the beam because of the force induced on the beam.
- You can change the contact between the surface of the beam and block to **Allow Penetration** using the **Contact Sets** button and check for the results if you are interested.
- To animate any of the result, right-click on the result name from the **Analysis Manager** and select the **Animate** button from the shortcut menu; refer to Figure-31.

Figure-31. Animate button

- A simulation will be played for the current selected result.

PERFORMING NONLINEAR STATIC ANALYSIS ON A RING UNDER APPLICATION OF TORQUE

The model used in this example is a ring of cast iron under the load of a torque. The steps to perform the analysis are given next.

Starting Non-Linear Static Analysis

- Open the part on which you want to perform the analysis.
- Click on the **SolidWorks Simulation** button from the **SOLIDWORKS Add-Ins** tab of the **Ribbon** to add **Simulation** tab in the **Ribbon**, if not added already.
- Click on the Down arrow below the **Study Advisor** button and select the **New Study** tool from the drop-down. List of analysis studies that can be performed, will be displayed in the left.
- Click on the **Nonlinear** button and then on the **Static** button from the **Options** rollout.
- Click on the **OK** button from the **PropertyManager**. The tools related to **Non-linear Static** analysis will be displayed.

Applying the Material

- Click on the **Apply Material** button from the **Ribbon**. The **Material** dialog box will be displayed.
- Select the **Alloy Steel** material from the left and click on the **Apply** button.
- Click on the **Close** button from the dialog box. The material will be applied.

Applying Fixtures

- Click on the down arrow below **Fixtures Advisor** in the **Ribbon**. The tools related to fixtures will be displayed.
- Click on the **Fixed Geometry** button from the tool list. The **Fixture PropertyManager** will be displayed.
- Select the faces as shown in Figure-32. Click on the **OK** button to apply fixture.

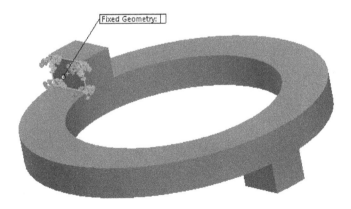

Figure-32. Face selected for fixture

- Since this is a round part and can deviate from its axis, so we need to restrict movement of its axis.

- Click again on the down arrow below the **Fixtures Advisor** button and select the **Fixed Hinge** button from the tool list. The **Fixture PropertyManager** will be displayed; refer to Figure-33.

Figure-33. Fixture Property Manager

- Select the inner round face of the model to make a fixed hinge; refer to Figure-34.

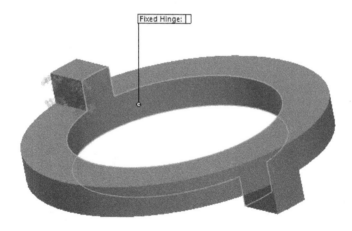

Figure-34. Round face selected for fixed hinge

- Click on the **OK** button from the **PropertyManager** to create the fixed hinge.

Applying Forces

- Click on the down arrow below **External Loads Advisor** button in the **Ribbon**. List of the tools related to loads will be displayed.
- Click on the **Torque** button from the list. The **Force/Torque PropertyManager** will be displayed.

- Note that **Variation with Time** rollout is added in the **PropertyManager** and two radio buttons are displayed named, **Linear** and **Curve**. Make sure that the **Linear** radio button is selected in the rollout.
- Enter the value of force as **1000** N in the **Force Value** edit box.
- Select the face as shown in Figure-35.

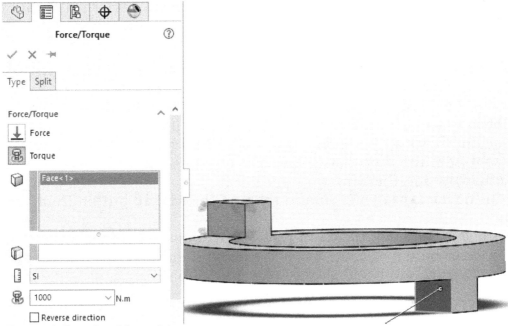

Figure-35. Face selected for applying torque

- Click in the pink box in **PropertyManager** and select the inner round face that was selected for fixed hinge. Preview of the torque will be selected.
- If change of torque direction is required then click on the **Reverse direction** check box. Click on the **OK** button from the **PropertyManager** to apply the torque.

Running the analysis

- Click on the **Run** button from the **Ribbon**. The analysis will start.
- After the processing is complete, the results will be displayed in the modeling area; refer to Figure-36.

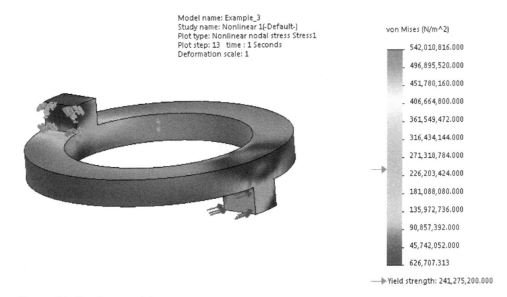

Figure-36. Result on applying torque

2D SIMPLIFICATION

2D Simplification is a method of converting simple 3D models to 2D geometries so that analysis can be performed faster. **2D Simplification** option is available for Static, Non-Linear Static, Non-Linear Dynamic, and Thermal analysis. In this chapter, we will discuss the Non-Linear Static analysis using 2D simplification. The same method is also applicable for other analyses. The procedure to perform non-linear static analysis using 2D simplification is given next.

Starting Non-Linear Static Analysis with 2D Simplification

- Open the part on which you want to perform the analysis.
- Click on the **SolidWorks Simulation** button from the **SOLIDWORKS Add-Ins** tab of the **Ribbon** to add **Simulation** tab in the **Ribbon**, if not added already.
- Click on the Down arrow below the **Study Advisor** button and select the **New Study** tool from the drop-down. List of analysis studies that can be performed, will be displayed in the left.
- Click on the **Nonlinear** button and then on the **Static** button from the **Options** rollout.
- Select the **Use 2D Simplification** check box from the **Options** rollout.
- Click on the **OK** button from the **PropertyManager**. The **Nonlinear (2D Simplification) PropertyManager** will be displayed; refer to Figure-37.

Figure-37. Nonlinear (2D Simplification) PropertyManager

- There are four buttons in the **Study Type** rollout of the **PropertyManager**; **Plane Stress**, **Plane Strain**, **Extrude** and **Axi-symmetric**.

  - Select the **Plane Stress** button if you want to analyze thin geometries with one dimension larger than the others. Note that the stress acting normal to the section plane is assumed as zero or negligible. This button is not available for thermal study.

- Select the **Plane Strain** button if you want to analyze with thick geometry assuming strain perpendicular to section plane is zero or negligible. This button is not available for thermal study.

- Select the **Extrude** button if thermal load is constant along the extrusion direction. This button is available for thermal study only.

- Select the **Axi-symmetric** button when the geometry, material properties, structural and thermal loads, fixtures, and contact conditions are symmetric (360^0) about an axis.

- If you have selected the **Plane Stress** or **Plane Strain** button then select the section plane and enter the desired section depth; refer to Figure-38. If you have selected the **Axi-symmetric** button then select the section plane and axis about which the part is symmetric; refer to Figure-39.

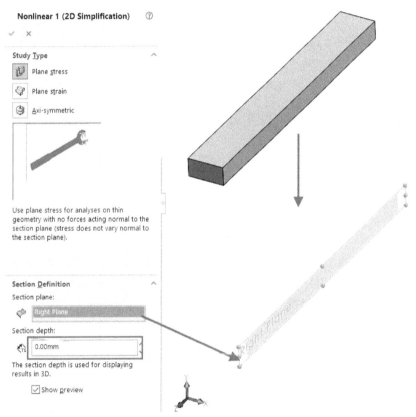

Figure-38. 2D simplification using Plane stress button

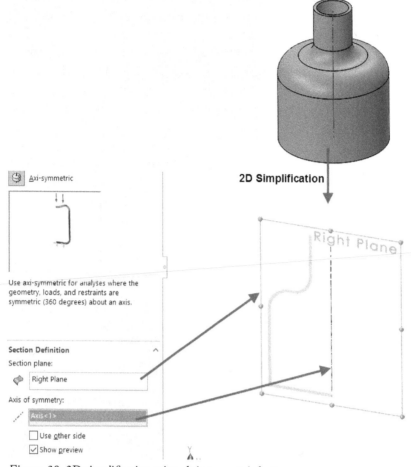

Figure-39. 2D simplification using Axi-symmetric button

- Click on the **OK** button from the **PropertyManager** after selecting desired options. The 2D simplified geometry of the model will be displayed; refer to Figure-40.

*Figure-40. Axi-symmetric 2D simpli-
fied geometry created*

Apply Boundary Conditions

- Apply material, fixture and other boundary conditions as required; refer to Figure-41.

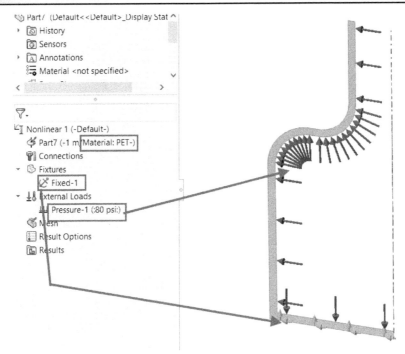

Figure-41. Boundary conditions for 2D model

Interpreting Results

- Run the analysis by using **Run This Study** button. In a few moments, you will get the result of analysis; refer to Figure-42.

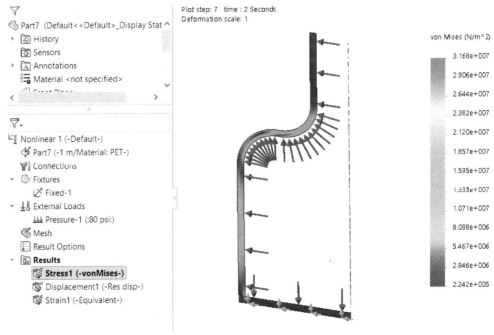

Figure-42. Result of analysis on 2D simplified model

- Note that the result is displayed in 2D form but if you want to check the results in 3D then right-click on Stress result and select the **Show as 3D Plot** option; refer to Figure-43.

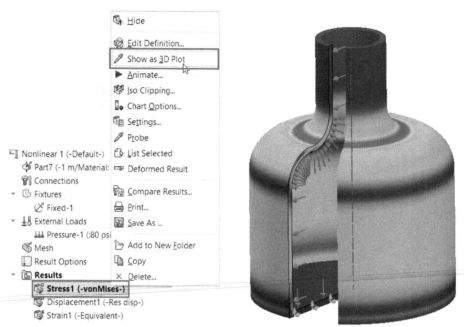

Nonlinear 1 (-Default-)
 Part7 (-1 m/Material:
 Connections
▼ Fixtures
  Fixed-1
▼ External Loads
  Pressure-1 (:80 psi
 Mesh
 Result Options
▼ **Results**
  Stress1 (-vonMises-)
  Displacement1 (-Res disp-)
  Strain1 (-Equivalent-)

Figure-43. Show as 3D Plot option

Figure-44. Stress plot in 3D

PRACTICE 1

Check out what happens to the fork under the conditions specified in Figure-45. Note that the material of fork is **Alloy Steel** and it is in the hands of a nasty kid. (You know kids!! They don't exert linear forces.)

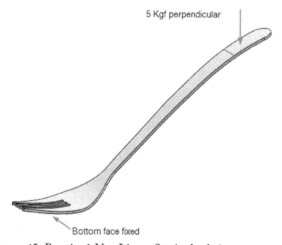

Figure-45. Practice 1 Non Linear Static Analysis

PRACTICE 2

Find out the deflection that will occur in the rope if 70 kg load is applied on the pin in the pulley; refer to Figure-46. Also, analyze the pulley if 200 kg load is applied on the pin and the rope acts as fixed steel bar.

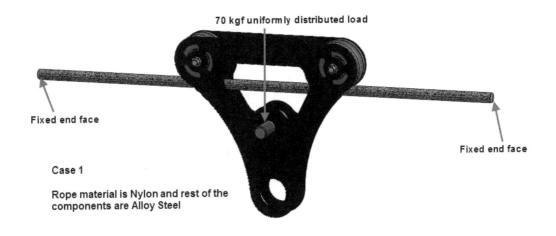

70 kgf uniformly distributed load

Fixed end face

Fixed end face

Case 1

Rope material is Nylon and rest of the components are Alloy Steel

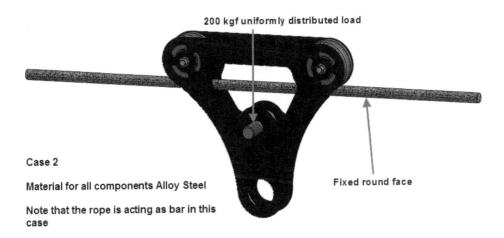

200 kgf uniformly distributed load

Case 2

Material for all components Alloy Steel

Note that the rope is acting as bar in this case

Fixed round face

Figure–46. Practice 2 Non Linear Analysis

SELF-ASSESSMENT

Q1. Which of the following is generally not considered as cause of non-linearity in Non-linear static analysis?

a. Large displacement in body
b. Non-linearity of material
c. Changing contact conditions
d. Change in force

Q2. Select the .... check box in the **Nonlinear-static Properties** dialog box to use the formula of large displacement.

Q3. If you want to use pre-stresses of SolidWorks Plastics then click on the .... tab in the **Nonlinear-static Properties** dialog box.

Q4. The equation for linear static analysis is given as ....

Q5. The equation for Newton-Raphson Method of non-linear analysis is given as ....

Q6. In context of Non-linear analysis, multiple sub-steps are used to apply the load gradually and each sub-step represents one load increment. (T/F)

Q7. You can see the intermediate results in non-linear analysis if you have selected the **Show intermediate results up to current step** check box in the **Non linear analysis solver** dialog box. (T/F)

FOR STUDENT NOTES

FOR STUDENT NOTES

FOR STUDENT NOTES

Answer to Self Assessment :
1. c, 2. Use large displacement formulation, 3. In-mold stresses, 4. $[K(D)]\{D\} = \{F\}$, 5. $[K_T]\{\Delta u\} = \{F\} - \{F^{nr}\}$, 6. T, 7. T

Chapter 6

Non-Linear
Dynamic Analysis

Topics Covered

The major topics covered in this chapter are:

- *Introduction*
- *Starting Non-Linear Dynamic Analysis.*
- *Understanding Iteration techniques and Integration methods*
- *Applying Material.*
- *Defining Fixtures*
- *Applying loads*
- *Defining Connections*
- *Simulating Analysis*
- *Interpreting results*
- *Response Graphs*

NON LINEAR DYNAMIC ANALYSIS

In linear static analysis, the loads are applied gradually and slowly until they reach their full magnitude. After reaching their full magnitude, the loads remain constant (time-invariant). The accelerations and velocities of the excited system are negligible, therefore, no inertial and damping forces are considered in the formulation:

$$[K]\{u\} = \{f\}$$

where:

[K]: stiffness matrix

{u}: displacement vector

{f}: load vector

The solution produces displacements and stresses that are constant.

In linear dynamic analysis, the applied loads are time-dependent. The loads can be deterministic (periodic, non-periodic), or non-deterministic which means that they cannot be precisely predicted but they can be described statistically. The accelerations and velocities of the excited system are significant, therefore, inertial and damping forces should be considered in the formulation:

$$[M]\{\ddot{u}(t)\} + [C]\{\dot{u}(t)\} + [K]\{u(t)\} = \{f(t)\}$$

where:

[K]: stiffness matrix

[C]: damping matrix

[M]: mass matrix

{u(t)}: time varying displacement vector

$\{\ddot{u}(t)\}$: time varying acceleration vector

$\{\dot{u}(t)\}$: time varying velocity vector

{f(t)}: time varying load vector

The response of the system is given in terms of time histories (amplitudes versus time), or in terms of frequency spectra (peak values versus frequency).

For linear dynamic analysis, the mass, stiffness, and damping matrices do not vary with time. Can you guess for what type of analysis they change with the time? The answer is Non-linear Dynamic Analysis.

Preparing Part and Starting Nonlinear Dynamic Analysis

- Open the Windshield ball.SLDASM file from resource kit on which the analysis is to be performed. In this file, you will see a model of ball striking to the model of windshield. Note that a major section of windshield is not going to play role in the analysis. So it is better to remove the extra section of windshield by using the modeling tools.

- Click on the **Extrude Cut** tool from the **Assembly Features** drop-down of the **Assembly CommandManager** in the **Ribbon**; refer to Figure-1. You will be asked to select the sketching plane.

Figure-1. Extrude Cut tool

- Click on the **FeatureManager Design Tree** button in the left pane and select the **Top Plane**; refer to Figure-2. The tools for sketching will be activated in the **Sketch CommandManager**. Press **CTRL+8**(not from the numpad) from the keyboard to make the sketching plane parallel to screen.

Figure-2. Selecting Top plane

- Create a rectangle cutting half of the windshield; refer to Figure-3. Click on the **Exit Sketch** button at the top-left corner from the **Sketch CommandManager**.

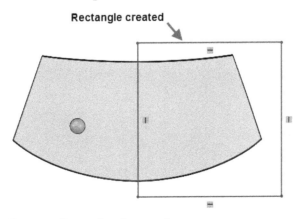

Rectangle created

Figure-3. Rectangle to be created

- Select the **Through All** option from the **Direction 1** rollout in the **Cut-Extrude PropertyManager** and click on the **OK** button from the **PropertyManager**. The model will be displayed as shown in Figure-4.
- Click on the **SolidWorks Simulation** button from the **SOLIDWORKS Add-Ins** tab of the **Ribbon** to add **Simulation** tab in the **Ribbon**, if not added already.
- Click on the Down arrow below the **Study Advisor** button and select the **New Study** tool from the drop-down. List of analysis studies that can be performed, will be displayed in the left.
- Click on the **Nonlinear** button and then on the **Dynamic** button from the **Options** rollout; refer to Figure-5.

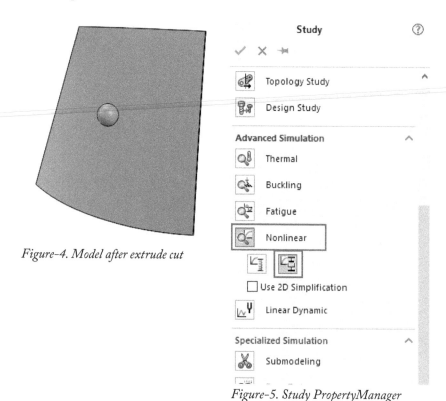

Figure-4. Model after extrude cut

Figure-5. Study PropertyManager

- Click on the **OK** button from the **PropertyManager**. The tools related to **Nonlinear Dynamic** analysis will be displayed. Note that most of the tools are same as discussed in previous chapters.

We know that the Nonlinear Dynamic analysis is the study of deformation as well as kinematics. Here, we have an example of car windshield (made of Acrylic plastic) and a steel ball; refer to Figure-6. Let's see what happens. (Is it National Geographic's Destroyed in seconds or not!!)

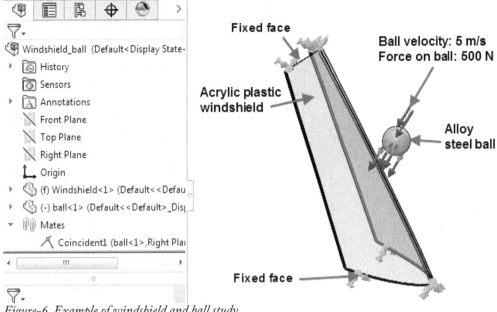

Figure-6. Example of windshield and ball study

Applying Material

- Right-click on **Windshield** under the **Parts** category in **Analysis Manager** and click on the **Apply/Edit Material** option from the shortcut menu. The **Material** dialog box will be displayed.

- Select the desired material, which is **Acrylic (Medium-high impact)** in the **Plastic** category from the left list in the dialog box and click on the **Apply** button.

- Click on the **Close** button from the dialog box to exit. Similarly, apply the **Alloy Steel** material to ball in the **Steel** category. The colors of models will be changed; refer to Figure-7.

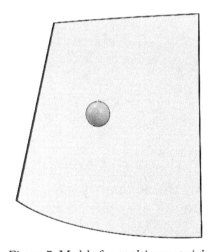

Figure-7. Model after applying material

Applying Fixtures

- Click on the down arrow below **Fixtures Advisor** in the **Ribbon**. The tools related to fixtures will be displayed.

- Click on the **Fixed Geometry** button from the tool list. The **Fixture PropertyManager** will be displayed; refer to Figure-8.

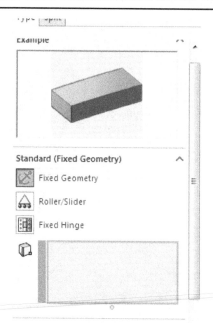

Figure-8. Fixture PropertyManager

- Select top and bottom flat faces of the windshield; refer to Figure-9.

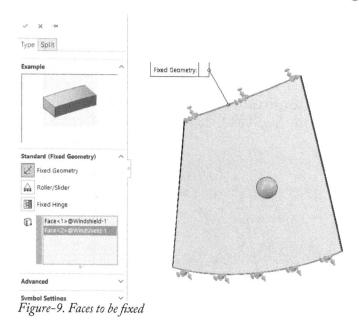

Figure-9. Faces to be fixed

- Click on the **OK** button from the **PropertyManager**. The boundary of windshield will be fixed as if they are bolted at some position.

Applying No Penetration Contact

- To apply no penetration contact between the components, we need to remove the global contact which is bonded by default. To do so, right-click on the **Component Contacts** option from the **Connections** category in the **Analysis Manager**. A shortcut menu will be displayed; refer to Figure-10.

Figure-10. Delete option for contacts

- Select the **Delete** option to delete the default contact. A warning message will be displayed. Click on the **Yes** button from the dialog box.
- Click on the **Contact Set** tool from the **Connections Advisor** drop-down in the **Ribbon**. The **Contact Sets PropertyManager** will be displayed.
- Click on the front faces of the Windshield part from the viewport and then click in the selection box for second set.
- Click on the round face of the ball; refer to Figure-11.

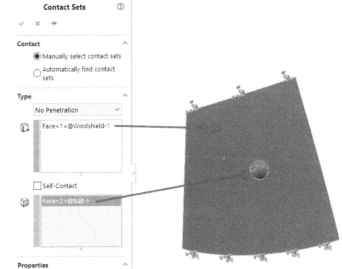
Figure-11. Faces selected for contact set

- Click on the **OK** button from the **PropertyManager** to apply contact.

Setting Initial Conditions

- Click on the **Initial Condition** tool from the **Ribbon**. The **Initial Conditions PropertyManager** will be displayed; refer to Figure-12.

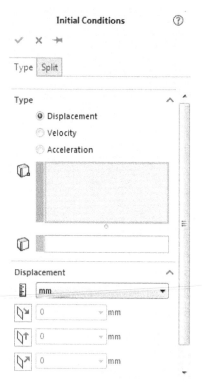

Figure-12. Initial Conditions PropertyManager

- Select the **Velocity** radio button from the **PropertyManager** to apply initial velocity to ball.
- Click on the ball part from **FeatureManager Design Tree** in the viewport.
- Click in the next selection box used for direction and select the Top plane of assembly for direction; refer to Figure-13.

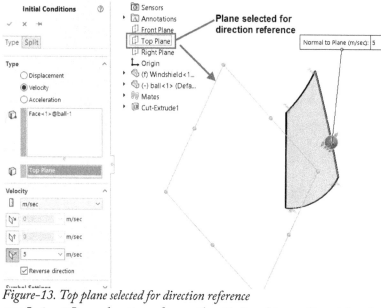

Figure-13. Top plane selected for direction reference

- Select the **Normal to Plane** button from the **Velocity** rollout in the **PropertyManager** and specify the value as **5** in the corresponding edit box with **Reverse direction** check box selected; refer to Figure-14. Note that to check the direction of force, you need to hover the cursor on the ball.

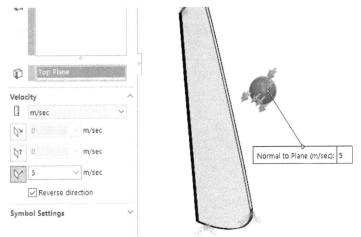

Figure-14. Velocity specified for ball

- Click on the **OK** button from the **PropertyManager** to apply initial condition.

Global Damping

Global Damping option is available in all the dynamic analyses in SolidWorks Simulation since we need to specify the damping coefficient for the study. Now, assume that the system on which we are going to perform the study is submerged in oil. To let SolidWorks Simulation understand that our system is submerged in a kind of oil, we need to specify global damping matrix. The steps to do so are given next.

- Click on the **Global Damping** button from the **Ribbon**. The **Rayleigh damping PropertyManager** will be displayed; refer to Figure-15.

Figure-15. Rayleigh damping PropertyManager

- Specify the damping parameters as required and click on the **OK** button. Note that for this particular problem, we do not have any damping conditions specified so we will not apply damping.

Rayleigh Damping

Rayleigh damping has certain mathematical conveniences and is widely used to model internal structural damping. The damping matrix C is given by $C = \alpha M + \beta K$ where M, K are the mass and stiffness matrices respectively and α, β are constants of proportionality. One of the less attractive features of Rayleigh damping is that the achieved damping ratio varies as response frequency varies. The stiffness proportional term contributes damping that is linearly proportional to response frequency and the mass proportional term contributes damping that is inversely proportional to response frequency. Mathematically, these frequency dependencies can be seen in the formula for damping ratio $\xi = \pi(\alpha/f + \beta f)$ where f is the response frequency.

The plot in Figure-16 illustrates how the separate mass and stiffness damping terms contribute to the overall damping ratio (Value of α is 0.025 and β is 0.023):

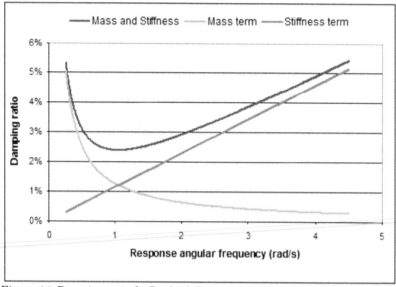

Figure-16. Damping ratio for Rayleigh Damping

Applying Force

- Click on the **Force** tool from the **External Loads Advisor** drop-down in the **Ribbon**. The **Force/Torque PropertyManager** will be displayed.
- Select the round face of ball and select the **Selected direction** radio button from the **PropertyManager**. The selection box for direction reference will be displayed.
- Select the Top Plane of assembly again.
- Click on the **Normal to Plane** button from the **Force PropertyManager** and specify the value as **500N** in the corresponding edit box with **Reverse direction** check box selected; refer to Figure-17.

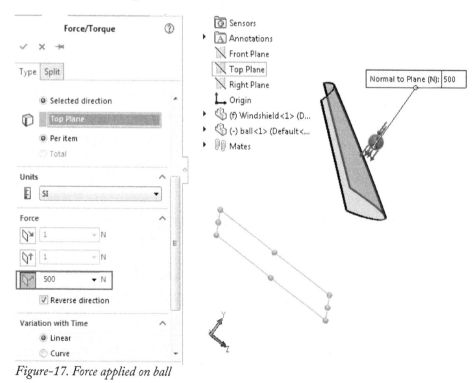

Figure-17. Force applied on ball

- Click on the **OK** button from the **PropertyManager** to apply the force.

Running the analysis

- Click on the **Run This Study** button from the **Ribbon**. The system will start analyzing the problem. Once the system is done, the results will be displayed in the **Modeling** area. Note that for analysis cases like this, you need to stop the analysis at the point where collision has occurred so that more resources are not wasted on the analysis which does not produce useful result; refer to Figure-18. After pausing analysis at desired time step, click on the **Cancel** button from the dialog box and click on the **Yes** button from the **Stop Solver** dialog box displayed. The analysis results will be displayed; refer to Figure-19.

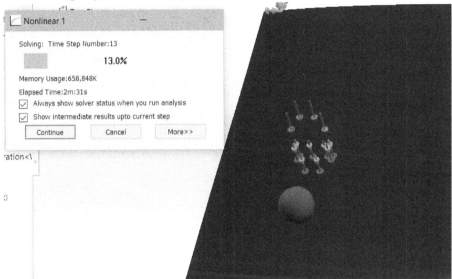

Figure-18. Analysis_stopped_at_intermediate_result.png

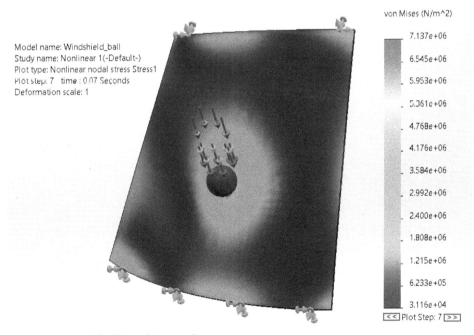

Figure-19. Windshield_analysis_result.png

Types of Results for Non Linear Dynamic Analysis

When you run an analysis on an assembly where effect of analysis are different on each of the components of assembly. So, you might want to check result of stress, strain, or deformation on any one of the assembly components. Here, we will check the vonMises stress generated on windshield only and remove the effects on ball.

VonMises Stress Report on Single Component of Assembly

*   After running analysis, click on the **Stress** option from **New** cascading menu in the **Results Advisor** drop-down of the **Ribbon**. The **Stress Plot PropertyManager** will be displayed.
*   Expand the **Advanced Options** rollout in the **PropertyManager** and select the **Show plot only on selected entities** check box. You will be asked to select the faces for displaying plot and the options in **PropertyManager** will be displayed as shown in Figure-20.

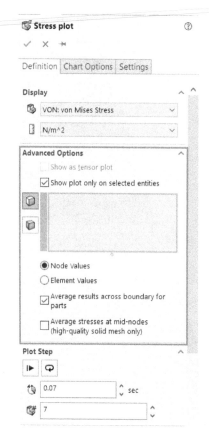

Figure-20. Advanced_options_of_stress_plot

*   By default, the **Faces** button is selected in the **Advanced Options** rollout. Select the **Bodies** button and select the body to be used for plotting result; refer to Figure-21.

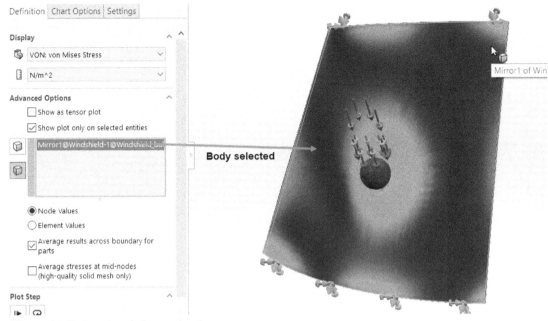

Figure-21. Body_selected_for_result_plot.png

- Set the desired time step and other parameters for the plot, and click on the **OK** button. The stress plot for single component of the assembly will be displayed.

Similarly, you can plot velocity and acceleration of ball in this case. The procedure to create Time History graph is discussed later in this chapter.

You can interpret the results as discussed in earlier chapters.

SETTING PARAMETERS IN NONLINEAR DYNAMIC ANALYSIS

As discussed in previous chapter, the method to solve nonlinear equation is Newton-Raphson method which is applicable for nonlinear dynamic analyses also. Before we start analyzing the model, we are required to set a few options in **Properties** dialog box. The procedure to do so is given next.

- After starting the nonlinear dynamic analysis, click on **Study Properties** button from the **Study Advisor** drop-down in the **Ribbon**. The **Nonlinear - Dynamic** dialog box will be displayed; refer to Figure-22.

Most of the options in the dialog box are same as discussed in previous chapters. We can set the time steps as discussed earlier. Here, we will discuss the advanced options which have not been discussed in previous chapters.

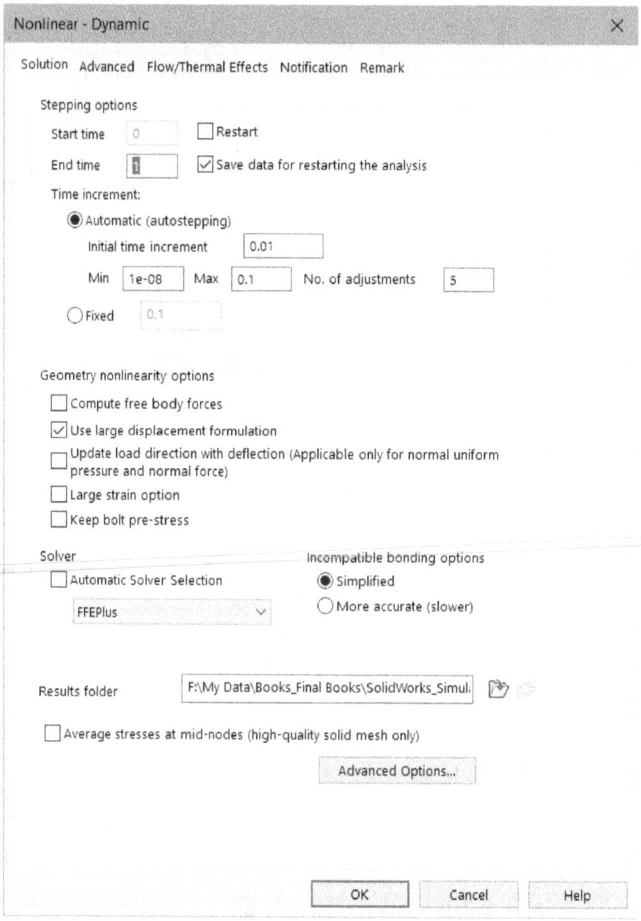

Figure-22. Nonlinear-Dynamic dialog box

- Click on the **Advanced Options** button from the dialog box. The **Advanced** tab will be added in the dialog box; refer to Figure-23.

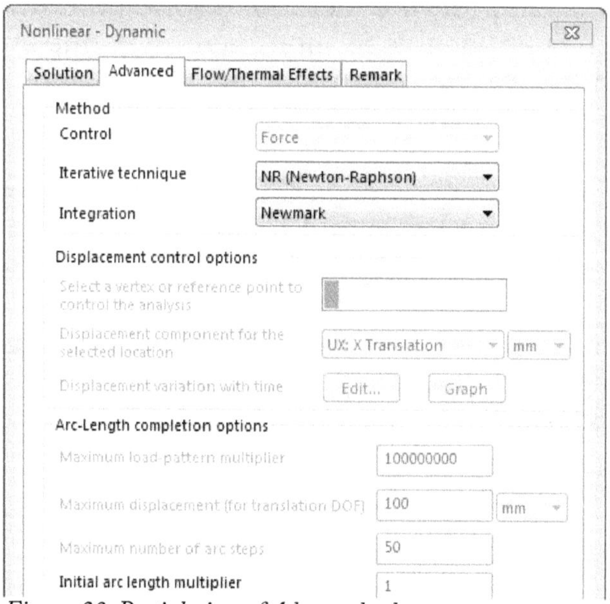

Figure-23. Partial view of Advanced tab

- Click in the **Iterative technique** drop-down and select the desired technique.

There are two options in the drop-down; **Newton-Raphson** and **MNR** (**Modified Newton-Raphson**). The Newton-Raphson method solves equation up to first derivative of function whereas modified Newton-Raphson method solves equation up to second derivative of function. More details of these options are given in next topic.

- Click in the drop-down for **Integration** and select the desired option to use as integration method. There are three options in this drop-down; Newmark, Wilson-Theta, and Central Difference. The details of these options is given after next topic in this chapter.

- Set the convergence tolerance and step parameters and then click on the **OK** button from the dialog box to apply settings.

Newton-Raphson and Modified N-R Method

For Newton-Raphson Method :

$$x_{i+1} = x_i + \frac{f(x_i)}{f'(x_i)}$$

For Modified Newton-Raphson Method:

$$x_{i+1} = x_i - \frac{f(x_i)f'(x_i)}{[f'(x_i)]^2 - f(x_i)f''(x_i)}$$

From the equations, we can understand that the Modified Newton-Raphson method is faster for multiple root equation because it tells slope as well as curvature of the function. But for simpler equations in which we need to find out one or two roots it takes more processing than the original Newton-Raphson method. So, we suggest the use of Newton-Raphson method when you have simpler problems and for complex problems, you should use Modified Newton-Raphson method as iterative technique.

Integration Method

Although there are a few more integration methods but we will concentrate on the three methods named earlier since these methods are used by SolidWorks Simulation.

Newmark Method

Newmark is a type of implicit method for integration which means differential equations are combined with equations of motions and the displacements are calculated by directly solving the equation. The Newmark method also assumes that the acceleration varies linearly between two instants of time. The equation for velocity and displacement in Newmark method can be given by Figure-24.

$$\dot{X}_{t+\Delta t} = \dot{X}_t + [(1-\alpha)\ddot{X}_t + a\ddot{X}_{t+\Delta t}]\,\Delta t$$

$$X_{t+\Delta t} = X_t + \dot{X}_t\,\Delta t + \left[(\frac{1}{2}-\beta)\ddot{X}_t + \beta\ddot{X}_{t+\Delta t}\right]\Delta t^2$$

Figure-24. Equation for Newmark Beta method

The parameters alpha and beta are user-defined and can be changed to suit the requirements of particular problems. For example, putting $\alpha = 1/2$ and $\beta = 0$ makes the acceleration constant and equal to $\ddot{X}_t$. We know the equation of force as,

$$\left\lfloor \overline{M} \right\rfloor \{X_{t+\Delta t}\} = \{\overline{F}_{t+\Delta t}\}$$

Here [M] is mass matrix and $\{F_{t+\Delta t}\}$ is force matrix

The Equations for Mass and Force can be given by,

$$[\overline{M}] = \frac{1}{\beta \Delta t^2}[M] + \frac{\alpha}{\beta \Delta t}[C] + [K]$$

$$\{\overline{F}_{t+\Delta t}\} = \{F_{t+\Delta t}\} + \left[(\frac{1}{2\beta} - 1)[M] + \Delta t(\frac{\alpha}{2\beta} - 1)[C] \right]\{\ddot{X}_t\}$$

$$+ \left[\frac{1}{\beta \Delta t}[M] + (\frac{\alpha}{\beta} - 1)[C] \right]\{\dot{X}_t\}$$

$$+ \left[\frac{1}{\beta \Delta t^2}[M] + \frac{\alpha}{\beta \Delta t}[C] \right]\{X_t\}$$

Some of the important points for using Newmark beta method are:
* In Newmark method, the amplitude in linear systems is conserved and the response is unconditionally stable if

$$\alpha \geq \frac{1}{2} \text{ and } \beta \geq \frac{1}{4}(\alpha + \frac{1}{2})^2$$

* If alpha = 1/2 and beta = 1/4 then large truncation errors occur in frequency of response.
* If beta = 1/2 then spurious damping is introduced proportional to (alpha-1/2).
* If beta is negative then self-excited vibrations arouse and if beta is greater than 1/2, a positive damping is introduced.

In straight words, the selection of alpha and beta can make big difference in results of analysis so their value should be decided very carefully.

Algorithm based on Newmark method
a. Initial Computations:
  1. Form stiffness [K], mass [M], and damping [C] matrices.

2. Initialize $\{X_0\}, \{\dot{X}_0\}$, and $\{\ddot{X}_0\}$.
3. Select the time step Δt, parameters α & β and calculate integration constants, $\beta \geq 0.5$ and $\alpha \geq 0.25(0.5+\beta)^2$.

$$a_0 = \frac{1}{\beta(\Delta t)^2}; \ a_1 = \frac{\alpha}{\beta \Delta t}; \ a_2 = \frac{1}{\beta \Delta t}; \ a_3 = \frac{1}{2\beta}; \ a_3 = \frac{1}{2\beta} - 1; \ a_4 = \frac{\alpha}{\beta} - 1;$$

$$a_5 = \frac{\Delta t}{2}\left(\frac{\alpha}{\beta} - 2\right); \ a_6 = \Delta t \, (1 - \beta); \ a_7 = \beta \Delta t$$

4. Form effective stiffness matrix:

$$\lfloor \overline{K} \rfloor = [K] + a_0 \, [M] + a_1 \, [C]$$

5. Triangularize $[\overline{K}]$: $[\overline{K}] = [L][D][L]^T$

b. For Each time step, calculate:

1. Calculate effective force vector at time $t + \Delta t$:

$$\{\overline{F}_{t+\Delta t}\} = \{F_{t+\Delta t}\} + [M](a_0\{X_t\} + a_2\{\dot{X}_t\} + a_2\{\dot{X}_t\} + a_3\{\ddot{X}_t\})$$

$$+ [C](a_1\{X_t\} + a_4\{\dot{X}_t\} + a_5\{\ddot{X}_t\})$$

2. Solve for displacements at time $t + \Delta t$

$$\lfloor \overline{K} \rfloor \{X_{t+\Delta t}\} = \{\overline{F}_{t+\Delta t}\}$$

3. Calculate $\{\dot{X}\}$ and $\{\ddot{X}\}$ at time $t + \Delta t$:

$$\{\ddot{X}_{t+\Delta t}\} = a_0\left(\{X_{t+\Delta t}\} - \{X_t\}\right) - a_2\{\dot{X}_t\} - a_3\{\ddot{X}_t\}$$

$$\{\dot{X}_{t+\Delta t}\} = a_1\left(\{X_{t+\Delta t}\} - \{X_t\}\right) - a_4\{\dot{X}_t\} - a_5\{\ddot{X}_t\}$$

Wilson-Theta Method

Wilson-Theta method is also a type of implicit method for integration. It assumes that the acceleration of system varies linearly between two instants of time $t_i = i\Delta t$ to $t_i + \theta\Delta t$, where $\theta \geq 1.0$. If τ is time increment between t and $t+\theta\Delta t$ then acceleration is given as,

$$\ddot{X}_{t+\tau} = \ddot{X}_t + \frac{\tau}{\theta\Delta t}\left(\ddot{X}_{t+\theta\Delta t} - \ddot{X}_t\right)$$

On successive integration, displacement is given by,

$$X_{t+\tau} = X_t + \dot{X}_t\,\tau + \frac{1}{2}\ddot{X}_t\,\tau^2 + \frac{\tau^3}{6\theta\Delta t}\left(\ddot{X}_{t+\theta\Delta t} - \ddot{X}_t\right)$$

Putting values in force equation we get,

$$[\overline{M}] = \frac{6}{\theta^2\Delta t^2}[M] + \frac{3}{\theta\Delta t}[C] + [K]$$

$$\{\overline{F}_{t+\theta\Delta t}\} = \{F_{t+\theta\Delta t}\} + (\frac{6}{\theta^2\Delta t^2}[M] + \frac{3}{\theta\Delta t}[C])\{X_t\}$$

$$+ (\frac{6}{\theta\Delta t}[M] + 2\,[C])\{\dot{X}_t\} + (2[M] + \frac{\theta\Delta t}{2}[C])\{\ddot{X}_t\}$$

Some of the important points for using Wilson theta method are:
- When θ=0 then method reduces to linear acceleration scheme.
- The method is unconditionally stable for θ≥1.37.
- In this method, there is no need to specify initial conditions because the displacement, velocity, and acceleration are functions of time.

Central Difference Method

Central Difference method is a type of explicit method for integration which means the response quantities are expressed in terms of previously determined values of displacement, velocity, and acceleration. It is assumed that the time period is divided into n equal parts of interval. The accuracy of solution depends on size of time step. The critical time step is given by $\Delta t_{cri} = \tau_n / \Pi$, where τ_n is natural period of system. If Δt is selected larger than Δt_{cri} then method becomes unstable making dynamic analysis meaningless due to big error. The equation of force for central difference method is given by,

$$M\left\{\frac{X_{i+1} - 2X_i + X_{i-1}}{(\Delta t)^2}\right\} + C\left\{\frac{X_{i+1} - X_{i-1}}{2\Delta t}\right\} + KX_i = F_i$$

On solving for displacement, we get

$$X_{i+1} = \left\{\frac{1}{\frac{M}{(\Delta t)^2} + \frac{C}{2\Delta t}}\right\}\left[\left\{\frac{2M}{(\Delta t)^2} - K\right\}\{X_i\} + \left\{\frac{C}{2\Delta t} - \frac{M}{(\Delta t)^2}\right\}X_{i-1} + F_i\right]$$

From the equation, you can easily interpret that the displacement of mass X_{i+1} can be calculated only if you have value of X_i, X_{i-1} and present force F_i. But initial conditions provide only X_0 and $\dot{X}_0$ so X_{-1} is calculated to find out X_1. First, the acceleration $\ddot{X}_0$ is calculated and then the value of X_{-1} is calculated by using it as given next,

$$\ddot{X}_0 = \frac{1}{M}[F(t=0) - C\dot{X}_0 - KX_0]$$

$$X_{-1} = X_0 - \Delta t\,\dot{X}_0 + \frac{(\Delta t)^2}{2}\,\ddot{X}_0$$

Note that the time step should be less than Δt_{cri} for using of this method otherwise the results will not be meaningful.

PRACTICAL 1

In this practical, we will find out response curve of the model under conditions given in Figure-25.

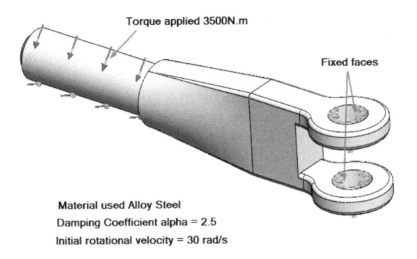

Figure-25. Practical 1 non linear dynamic

Starting Nonlinear Dynamic Analysis

- Open the part on which you want to perform the analysis.
- Click on the **SolidWorks Simulation** button from the **SOLIDWORKS Add-Ins** tab of the **Ribbon** to add **Simulation** tab in the **Ribbon**, if not added already.
- Click on the Down arrow below the **Study Advisor** button and select the **New Study** tool from the drop-down. List of analysis studies that can be performed, will be displayed in the left.
- Click on the **Nonlinear** button and then on the **Dynamic** button from the **Options** rollout; refer to Figure-26.

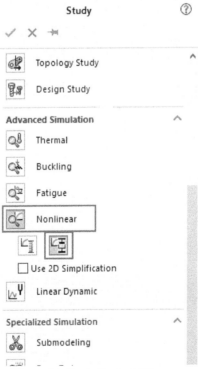

Figure-26. Study PropertyManager

- Click on the **OK** button from the **PropertyManager**. The tools related to **Nonlinear Dynamic** analysis will be displayed. Note that most of the tools are same as discussed in previous chapters.
- Select the desired method and integration technique for analysis. For this practical, we have used Newton-Raphson method with Newmark integration technique.

Applying Material

- Click on the **Apply Material** button from the **Ribbon**. The **Material** dialog box will be displayed as discussed in earlier chapters.
- Select the desired material, say **Alloy Steel** option from the left list and click on the **Apply** button.
- Click on the **Close** button from the dialog box to exit. The color of model will be changed; refer to Figure-27.

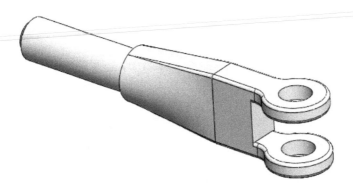

Figure-27. Model after applying material

Applying Time Varying Torque

- Click on the down arrow below **External Loads Advisor** button in the **Ribbon**. List of the tools related to loads will be displayed.
- Click on the **Torque** button from the list. The **Force/Torque PropertyManager** will be displayed.
- Select the round face of the model and specify the torque value as shown in Figure-28.

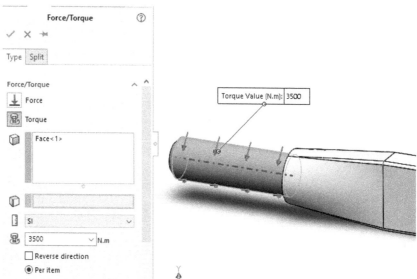

Figure-28. Torque value specified

- Expand the **Variation with Time** rollout in the **PropertyManager** and click on the **Curve** radio button to define the loading curve.
- Click on the **Edit** button from the **Variation with Time** rollout and define the time curve as shown in Figure-29. Note that to add a new row in the table, you need to double-click on the last field under **Points** column in the table.

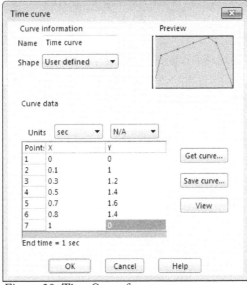

Figure-29. Time Curve for torque

- Click on the **OK** button from the dialog box and then **OK** button from the **PropertyManager**.

Applying Fixtures

- Click on the down arrow below **Fixtures Advisor** in the **Ribbon**. The tools related to fixtures will be displayed.
- Click on the **Fixed Geometry** button from the tool list. The **Fixture PropertyManager** will be displayed.
- Select the holes cut from the model as shown in Figure-30.

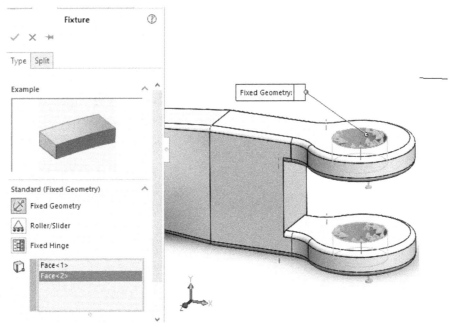

Figure-30. Holes fixed

• Click on the **OK** button from the **PropertyManager** to apply the fixtures.

Applying Damping

• Click on the **Global Damping** button from the **Ribbon**. The **Rayleigh damping PropertyManager** will be displayed.
• Specify the value of alpha as **2.5** in the corresponding edit box and then click on the **OK** button from the **PropertyManager**.

Specifying initial velocity

• Click on the **Initial Conditions** button from the **Ribbon**. The **Initial Conditions PropertyManager** will be displayed; refer to Figure-31.

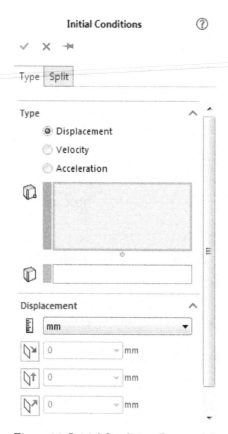

Figure-31. Initial Conditions PropertyManager

• Click on the **Velocity** radio button and select the round face of the model.
• Click in the pink selection box of **PropertyManager** to define the direction and then select the round face of the model again; refer to Figure-32.

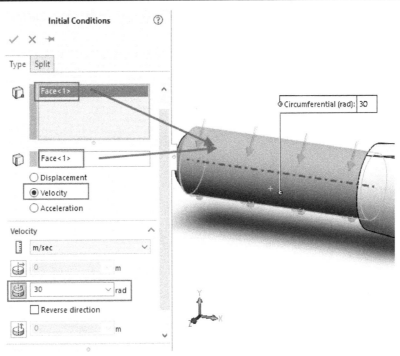

Figure-32. Face selected for defining velocity

- Select the **Circumferential** button from the **Velocity** rollout and specify the value as **30** in the corresponding edit box; refer to Figure-33.

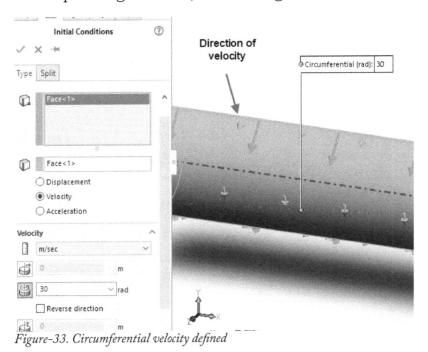

Figure-33. Circumferential velocity defined

- Click on the **OK** button from the **PropertyManager**.

Defining Mesh

Since we are performing the analysis for educational purpose only, here we will define the mesh as coarse. This will reduce the solving time of analysis.

- Click on the down arrow below **Run This Study** button in the **Ribbon**. List of tools will be displayed.

- Click on the **Create Mesh** tool from the list. The **Mesh PropertyManager** will be displayed.
- Move the slider to extreme left in the **Mesh Density** rollout in **PropertyManager**; refer to Figure-34.

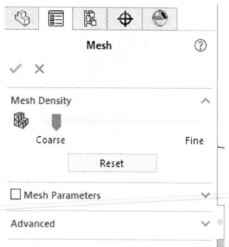

Figure-34. Coarse mesh density

- Click on the **OK** button from the **PropertyManager** to apply meshing.

Running analysis

- Click on the **Run This Study** button from the **Ribbon**. The intermediate results of analysis will be displayed; refer to Figure-35. Note that the intermediate results display is latest enhancement in SolidWorks Simulation.

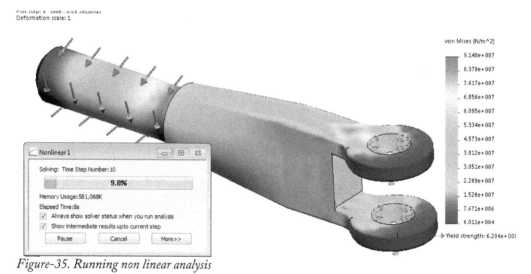

Figure-35. Running non linear analysis

Creating Time response curve

As discussed earlier, Time response curve is the main result for any Non-linear analysis because stress changes as per the force curve. It might be possible that the stress is very low at the end of study but very high in the middle of study. So, Time response curve gives the real time information of stress with respect to time. The steps to add Time response curve in results, are given next.

* Right-click on the **Result** node in the **Analysis Manager**. A shortcut menu will be displayed; refer to Figure-36.

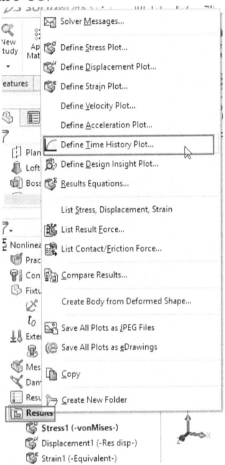

Figure-36. Shortcut menu for Time response curve

* Click on the **Define Time History Plot** option from the shortcut menu. The **Time History Graph PropertyManager** will be displayed; refer to Figure-37.
* Make sure that **Time** is selected in the **X axis** drop-down and **Stress** is selected in the **Y axis** drop-down as shown in Figure-37.

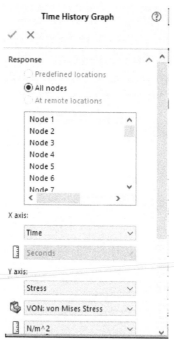

Figure-37. Time History Graph
PropertyManager

- Click on the **OK** button from the **PropertyManager**. The **Response Graph** will be displayed; refer to Figure-38. Note that this is the response graph of node 1 which was selected in the **Time History Graph PropertyManager** earlier; refer to previous figure (Figure-40).

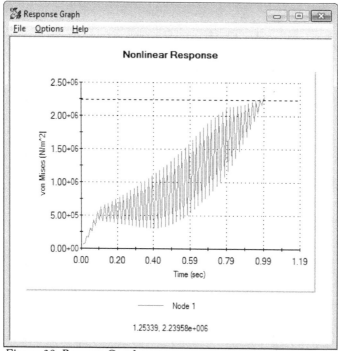

Figure-38. Response Graph

- Close the **Response Graph** dialog box and right-click on **Response1** result in the **Result** node in **Analysis Manager**; refer to Figure-39.

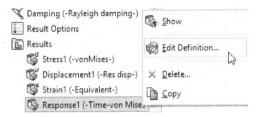

Figure-39. Shortcut menu for response result

- Select the **Edit Definition** option from the shortcut menu and select a different node to check the response. Like, select the node 95 (Figure-40) from the list and click **OK** from the **PropertyManager** to see the response graph(Figure-41).

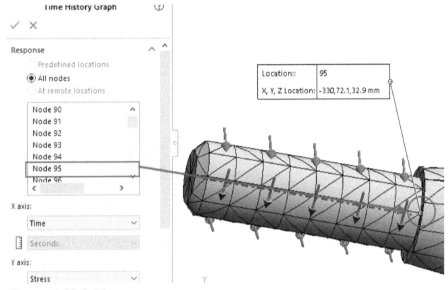

Figure-40. Node 95

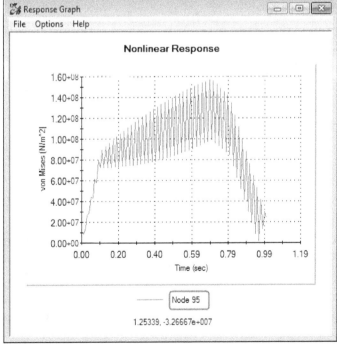

Figure-41. Response Graph for node 95

- In this way, you can check the response graph of critical areas. Note that the stress of this node has gone up to 1.566×10^8 but in final **Stress1** result, we don't find any spot with such high stress; refer to Figure-42. So, you can now understand why Response graph is important in case of non-linear analyses.

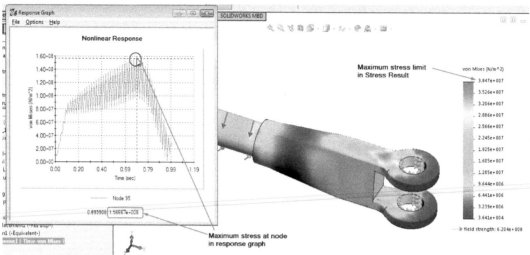

Figure-42. Comparing stress results

PRACTICE 1

A sheet metal plate of **140mmX40mmX3mm** is placed on bending machine with bending line at the center of plate along length. The bending tool has a radius of 3 mm. The load applied by bending machine on plate is **100 kgf**. Find out if plate will bend or not. Material is Stainless Steel.

SELF-ASSESSMENT

Q1. For linear dynamic analysis, the mass, stiffness, and damping matrices do not vary with time. (T/F)

Q2. Global Damping option is not available in all type of dynamic analyses in SolidWorks Simulation. (T/F)

Q3. Central Difference method is a type of explicit method of integration. (T/F)

Q4. In a Non-linear analysis stress changes as per the force curve. (T/F)

Q5. Which of the following condition is required for stable solution of Newmark Integration Method?

a. $\alpha \geq 0.5$ and $\beta \geq 0.25(\alpha+0.5)^2$
b. $\alpha \leq 0.5$ and $\beta \geq 0.25(\alpha+0.5)^2$
c. $\alpha \geq 0.5$ and $\beta \leq 0.25(\alpha+0.5)^2$
d. $\alpha \leq 0.5$ and $\beta \leq 0.25(\alpha+0.5)^2$

Q6. In all cases, the Modified Newton-Raphson method is faster because it tells slope as well as curvature of the function. (T/F)

Q7. Newmark is a type of implicit method for integration which means differential equations are combined with equations of motions and the displacements are calculated by directly solving the equation. (T/F)

FOR STUDENT NOTES

FOR STUDENT NOTES

Answer to Self-Assessment :
1. F, 2. F, 3. T, 4. T, 5. a, 6. F, 7. T

Chapter 7

Frequency Analysis

Topics Covered

The major topics covered in this chapter are:

- *Introduction*
- *Starting Frequency Analysis.*
- *Applying Material.*
- *Defining Fixtures*
- *Applying loads*
- *Defining Connections*
- *Simulating Analysis.*
- *Interpreting results*

INTRODUCTION

Every structure has the tendency to vibrate at certain frequencies, called **natural or resonant frequencies**. Each natural frequency is associated with a certain shape, called **mode shape**, that the model tends to assume when vibrating at that frequency.

When a structure is properly excited by a dynamic load with a frequency that coincides with one of its natural frequencies, the structure undergoes large displacements and stresses. This phenomenon is known as **resonance**. For undamped systems, resonance theoretically causes infinite motion. **Damping**, however, puts a limit on the response of the structures due to resonant loads.

A real model has an infinite number of natural frequencies. However, a finite element model has a finite number of natural frequencies that are equal to the number of degrees of freedom considered in the model. Only the first few modes are needed for most purposes.

If your design is subjected to dynamic environments, static studies cannot be used to evaluate the response. Frequency studies can help you design vibration isolation systems by avoiding resonance in specific frequency band. They also form the basis for evaluating the response of linear dynamic systems where the response of a system to a dynamic environment is assumed to be equal to the summation of the contributions of the modes considered in the analysis.

Note that resonance is desirable in the design of some devices. For example, resonance is required in guitars and violins.

The natural frequencies and corresponding mode shapes depend on the geometry, material properties, and support conditions. The computation of natural frequencies and mode shapes is known as modal, frequency, and normal mode analysis.

When building the geometry of a model, you usually create it based on the original (undeformed) shape of the model. Some loads, like the structure's own weight, are always present and can cause considerable effects on the shape of the structure and its modal properties. In many cases, this effect can be ignored because the induced deflections are small.

Loads affect the modal characteristics of a body. In general, compressive loads decrease resonant frequencies and tensile loads increase them. This fact is easily demonstrated by changing the tension on a violin string. The higher the tension, the higher the frequency (tone).

You do not need to define any loads for a frequency study but if you do, their effect will be considered. By having evaluated natural frequencies of a structure's vibrations at the design stage, you can optimize the structure with the goal of meeting the frequency vibro-stability condition. To increase natural frequencies, you would need to add rigidity to the structure and (or) reduce its weight. For example, in the case of a slender object, the rigidity can be increased by reducing the length and increasing the thickness of the object. To reduce a part's natural frequency, you should, on the contrary, increase the weight or reduce the object's rigidity.

Note that the software also considers thermal and fluid pressure effects for frequency studies. The procedure to perform the frequency analysis is given next.

STARTING FREQUENCY ANALYSIS

- Open the part on which you want to perform the analysis.
- Click on the **SolidWorks Simulation** button from the **SOLIDWORKS Add-Ins** tab of the **Ribbon** to add **Simulation** tab in the **Ribbon**, if not added already.
- Click on the Down arrow below the **Study Advisor** button and select the **New Study** tool from the drop-down. List of analysis studies that can be performed, will be displayed in the left.
- Click on the **Frequency** button then click on the **OK** button from the **PropertyManager**. The tools related to frequency analysis will be displayed.

Applying Material

- Click on the **Apply Material** button from the **Ribbon**. The **Material** dialog box will be displayed.
- Expand the **SolidWorks Materials** node and then the **Aluminium Alloys** node in the left of the dialog box.
- Browse through the materials and select the **Alumina** material.
- Click on the **Apply** button and then click on the **Close** button from the dialog box to exit the dialog box.

Applying Fixtures

- Click on the down arrow below **Fixtures Advisor** in the **Ribbon**. The tools related to fixtures will be displayed.
- Click on the **Fixed Geometry** button from the tool list. The **Fixture PropertyManager** will be displayed.
- Select the face of the model as shown in Figure-1.
- Click on the **OK** button from the **PropertyManager**.

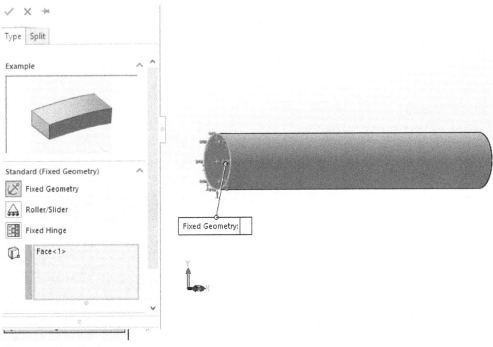

Figure-1. Face for fixture

Applying Forces

Note that it is not necessary to apply forces for performing the Frequency analysis but if you apply the forces, they are counted for their effects in deformation. They do not make any effect on resonance frequencies.

- Click on the down arrow below **External Loads Advisor** from the **Ribbon**. The list of tools will be displayed.
- Click on the **Force** button from the list. The **Force/Torque PropertyManager** will be displayed; refer to Figure-2.

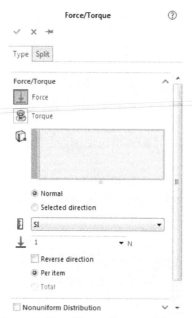

Figure-2. Force Torque PropertyManager

- Select the face as shown in Figure-3 and specify the force value as **100** N.

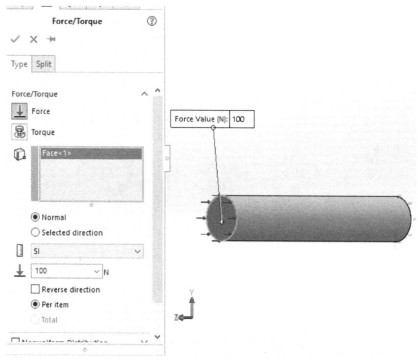

Figure-3. Force to be applied

- Click on the **OK** button from the **PropertyManager**.

Setting Parameters for the analysis

- Right-click on the name of analysis in the **Analysis Manager**. A shortcut menu will be displayed; refer to Figure-4.

Figure-4. Shortcut menu for properties

- Click on the **Properties** button from the shortcut menu. The **Frequency** dialog box will be displayed; refer to Figure-5.
- In this dialog box, the **Number of frequencies** spinner is used to specify the number of natural frequencies that are to be found out.
- If you want to check the natural frequency closer to a specific frequency then select the **Calculate frequencies closest to: (Frequency Shift)** check box and specify the value of frequency in the adjacent edit box.
- As we are applying force within the Frequency analysis, then we need to select the **Direct sparse Solver** from the **Solver** area in the dialog box. Otherwise, an error will be generated.

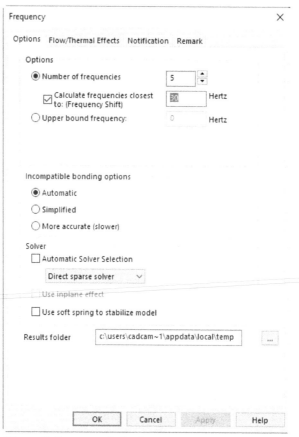

Figure-5. Frequency dialog box

- Other options in the dialog box have already been discussed.
- Click on the **OK** button from the dialog box.

Running the analysis

- Click on the **Run This Study** button from the **Ribbon** to run the analysis. The solving process of analysis will start.
- After the process is complete, the results will be displayed; refer to Figure-7.

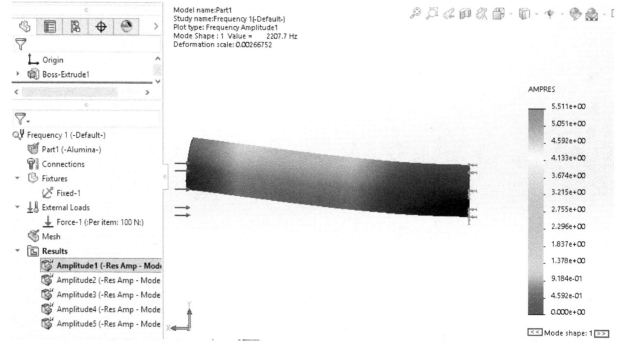

Figure-6. Result of Frequency analysis

- There are six nodes displayed under **Results** in the **Analysis Manager** which represent six mode shapes at respective natural frequencies.
- From the results, you find that our part should not be working in the range of the natural frequencies listed.

Results

- To display the list of resonant frequencies, right-click on the **Results** node in the **Analysis Manager**. A shortcut menu will be displayed.

Resonant Frequencies List

- Click on the **List Resonant Frequencies** button from the shortcut menu; refer to Figure-7.

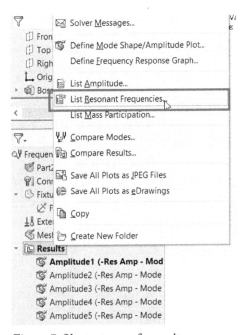

Figure-7. Shortcut menu for results

- The **List Modes** dialog box will be displayed. The **Frequency (Hz)** column shows the resonant frequencies; refer to Figure-8.

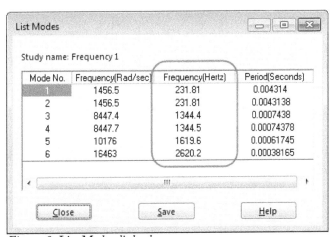

Figure-8. List Modes dialog box

Similarly, you can create Displacement and Mass participation plots for different frequencies using the related tools from the **Result Advisor** drop-down.

SELF-ASSESSMENT

Q1. Every structure has the tendency to vibrate at certain frequencies, called .... or .... frequencies.

Q2. Each natural frequency is associated with a certain shape, called....................

Q3. The structure undergoes large displacements and stresses under the load with frequencies coinciding with its natural frequencies. This phenomenon is known as ....

Q4. Frequency studies can help you design vibration isolation systems by avoiding resonance in specific frequency band. (T/F)

Q5. To increase a part's natural frequency, you should increase the weight or reduce the object's rigidity. (T/F)

Q6. For undamped systems, resonance theoretically causes infinite motion. (T/F)

Q7. A real model has an infinite number of natural frequencies. (T/F)

Q8. .... analysis can help you design vibration isolation systems by avoiding resonance in specific frequency band.

FOR STUDENT NOTES

FOR STUDENT NOTES

Answer to Self-Assessment :
1. natural or resonant, 2. mode shape, 3. resonance, 4. T, 5. F, 6. T, 7. T, 8. Frequency

Chapter 8

Linear Dynamic Analysis

Topics Covered

The major topics covered in this chapter are:

- *Introduction*
- *Types of Linear Dynamic Analysis*
- *Starting Linear Dynamic Analysis.*
- *Applying Material.*
- *Defining Fixtures*
- *Applying loads*
- *Defining Connections*
- *Simulating Analysis.*
- *Interpreting results*

LINEAR DYNAMIC ANALYSIS

Static studies assume that loads are constant or applied very slowly until they reach their full values. Because of this assumption, the velocity and acceleration of each particle of the model is assumed to be zero. As a result, static studies neglect inertial and damping forces.

For many practical cases, loads are not applied slowly or they change with time or frequency. For such cases, use a dynamic study. Generally if the frequency of a load is larger than 1/3 of the lowest (fundamental) frequency, a dynamic study should be used.

Linear dynamic studies are based on frequency studies. The software calculates the response of the model by accumulating the contribution of each mode to the loading environment. In most cases, only the lower modes contribute significantly to the response. The contribution of a mode depends on the load's frequency content, magnitude, direction, duration, and location.

Objectives of a dynamic analysis

• Design structural and mechanical systems to perform without failure in dynamic environments.
• Modify system's characteristics (i.e. geometry, damping mechanisms, material properties, etc.) to reduce vibration effects.

There are four type of dynamic analyses that can be performed in SolidWorks Simulation:

• Modal Time History Analysis
• Harmonic Analysis
• Random Vibration Analysis
• Response Spectrum Analysis

Before we start knowing about various dynamic analyses, it is important to understand the concept of damping.

Damping

If you apply initial conditions to a dynamic system, the system vibrates with decreasing amplitudes until it comes to rest. This phenomenon is called damping. Damping is a complex phenomenon that dissipates energy by many mechanisms like internal and external friction, thermal effects of cyclic elastic straining materials at the microscopic level, and air resistance.

It is difficult to describe dissipation mechanisms mathematically. Damping effects are usually represented by idealized mathematical formulations. For many cases, damping effects are adequately described by equivalent viscous dampers.

A viscous damper (or dashpot) generates a force that is proportional to velocity. A piston that can move freely inside a cylinder filled with a viscous fluid like oil is an example of a viscous damper.

The following types of damping are available:

- Modal Damping
- Rayleigh Damping
- Composite Modal Damping
- Concentrated Dampers. Defined between two locations (available for modal time history analysis).

We have discussed about Rayleigh Damping in previous chapters. Now, we will discuss about the other damping options.

Modal Damping

Modal damping is defined as a ratio of the critical damping C_c for each mode. Critical damping C_c is the least amount of damping that causes a system to return to its equilibrium position without oscillating.

The modal damping ratio can be determined accurately with proper field tests. The ratio varies from 0.01 for lightly damped systems to 0.15 or more for highly damped systems.

When experimental data is not available, use data from a similar class of systems to determine the damping properties. Smaller ratios are more conservative since higher ratios reduce vibration amplitudes. In general, neglecting damping leads to a conservative estimate of the system's response.

In mathematical terms, damping ratio is given by

$$\zeta = \frac{c}{c_c},$$
$$\zeta = \frac{\text{actual damping}}{\text{critical damping}},$$

Here, $C_c = 2m\omega_n$

Damping ratios for different systems and materials are given as:

System	Viscous Damping Ratio ζ (as percentages of critical damping)
Metals (in elastic range)	less than 0.01
Continuous metal structures	0.02 - 0.04
Metal structures with joints	0.03 - 0.07
Aluminum / steel transmission lines	~ 0.04
Small diameter piping systems	0.01 - 0.02
Large diameter piping systems	0.02 -0.03
Auto shock absorbers	~ 0.30
Rubber	0.05
Large buildings during earthquake	0.01 - 0.05
Prestressed concrete structures	0.02 -0.05
Reinforced concrete structures	0.04 -0.07

Material	Viscous Damping Ratio ζ (under approximately 20 °C)
Aluminum	~ 0.5 10^{-4}
Lead (pure)	~ 10^{-2}
Iron	1 to 3 10^{-4}
Copper (polycrystalline)	10^{-3}
Magnesium	~ 0.5 10^{-4}
Brass	< 0.5 10^{-3}
Nickel	< 0.5 10^{-3}
Silver	< 1.5 10^{-3}
Bismuth	~ 4 10^{-4}
Zinc	~ 1.5 10^{-4}
Tin	~ 10 10^{-4}

Composite Modal Damping

Composite modal damping allows the definition of the damping as a material property. Material damping ratio is defined in the **Properties** tab of the **Material** dialog box. The program uses this property to calculate equivalent modal damping ratio for each mode.

The composite modal damping is defined in terms of equivalent modal damping ratios as:

$$\beta_j = \{\varphi\}_j^T [\overline{M}]\{\varphi\}_j$$

where:

β_j = equivalent modal damping ratio of the jth mode

$\{\varphi\}_j^T$ = j^{th} normalized modal eigenvector

$[\overline{M}]$ = modified global mass matrix constructed from the product of the element's damping coefficient and its mass matrix.

Here, the questions rise: **Where to use Modal damping? Where to use Rayleigh Damping? Why the Composite Modal damping is required?**

The answer to these questions are not straight forward but they are dependent of many factors like material structure, mass properties, damping system used, and many other properties of the system.

- Rayleigh Damping scheme allows to use the flexibility of using systems that change damping force depending on mass matrix and stiffness matrix of model. If you expect changes to happen in mass or stiffness then you should use Rayleigh Damping scheme otherwise it is good to use Modal damping scheme because you have a direct value of damping for most of the materials.

- If there is requirement of calculating internal damping of part then Modal damping scheme is better than Rayleigh damping scheme.
- If the part has softer elements then Rayleigh Damping can product unrealistic results because of stiffness matrix part. So, you should use Modal damping.

The Composite Modal Damping uses the damping value specified in the material properties. In this way, it reduces the requirement of explicitly defining damping value for each object in assembly under the study.

TYPES OF DYNAMIC ANALYSIS

On the basis of the use, the all four analyses can be explained as follows:

Modal Time History Analysis

Use modal time history analysis when the variation of each load with time is known explicitly and you are interested in the response as a function of time.

Typical loads include:

- Shock (or pulse) loads
- General time-varying loads (periodic or non-periodic)
- Uniform base motion (displacement, velocity, or acceleration applied to all supports)
- Support motions (displacement, velocity, or acceleration applied to selected supports non-uniformly)
- Initial conditions (a finite displacement, velocity, or acceleration applied to a part or the whole model at time t =0).

The solution of the equations of motion for multi degree-of-freedom systems incorporates Modal analysis techniques.

The solution's accuracy can improve by using a smaller time step.

After running the study, you can view displacements, stresses, strains, reaction forces, etc. at different time steps, or you can graph results at specified locations versus time. If no locations are specified in Result Options, results at all nodes are saved.

Modal, Rayleigh, composite modal, and concentrated dampers are available for modal time history analysis.

Harmonic Analysis

Use harmonic analysis to calculate the peak steady state response due to harmonic loads or base excitations.

A harmonic load P is expressed as $P = A \sin(\omega t + \varphi)$ where: A is the amplitude, ω is the frequency, t is time, and φ is the phase angle. Sample harmonic loads of different frequencies w versus time are shown in Figure-1.

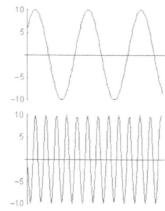

Figure-1. Graph for linear dynamic analysis

Although you can create a modal time history study and define loads as functions of time, you may not be interested in the transient variation of the response with time. In such cases, you save time and resources by solving for the steady-state peak response at the desired operational frequency range using harmonic analysis.

For example, a motor mounted on a test table transfers harmonic loads to the support system through the bolts. You can model the supporting system and define a harmonic study to evaluate the steady-state peak displacements, stresses, etc. for the motor's range of operating frequencies. You can approximate the motor by a distributed mass.

After running the study, you can view peak stresses, displacements, accelerations, and velocities as well as phase angles of the response over the range of operating frequencies.

Modal, Rayleigh, and Composite modal damping options are available for this type of analysis.

Random Vibration Analysis

Use a random vibration study to calculate the response due to non-deterministic loads. Examples of non-deterministic loads include:

*   Loads generated on a wheel of a car traveling on a rough road
*   Base accelerations generated by earthquakes
*   Pressure generated by air turbulence
*   Pressure from sea waves or strong wind

In a random vibration study, loads are described statistically by power spectral density (psd) functions. The units of psd are the units of the load squared over frequency as a function of frequency. For example, the units of a psd curve for pressure are $(psi)^2/Hz$.

The solution of random vibration problems is formulated in the frequency domain.

After running the study, you can plot root-mean-square (RMS) values, or psd results of stresses, displacements, velocities, etc. at a specific frequency or graph results at specific locations versus frequency values.

Modal, Rayleigh, and Composite modal damping options are available for this type of analysis.

Response Spectrum Analysis

In a response spectrum analysis, the results of a modal analysis are used in terms of a known spectrum to calculate displacements and stresses in the model. For each mode, a response is read from a design spectrum based on the modal frequency and a given damping ratio. All modal responses are then combined to provide an estimate of the total response of the structure.

You can use a response spectrum analysis rather than a time history analysis to estimate the response of structures to random or time-dependent loading environments such as earthquakes, wind loads, ocean wave loads, jet engine thrust or rocket motor vibrations.

After reading from the above section, we can find that the pattern of force application changes in all the analyses. Now, we will start with the Random Vibration Analysis and proceed to other Dynamic analyses.

Starting Random Vibration Analysis

- Open the part on which you want to perform the analysis.
- Click on the **SolidWorks Simulation** button from the **SOLIDWORKS Add-Ins** tab of the **Ribbon** to add **Simulation** tab in the **Ribbon**, if not added already.
- Click on the Down arrow below the **Study Advisor** button and select the **New Study** tool from the drop-down. List of analysis studies that can be performed, will be displayed in the left.
- Click on the **Linear Dynamic** button and then on the **Random Vibration Analysis** button from the **Options** rollout; refer to Figure-2.

Figure-2. Study PropertyManager

- Click on the **OK** button from the **PropertyManager**. The tools related to **Linear Dynamic Random Vibration** analysis will be displayed. Note that most of the tools are same as discussed in previous chapters.

Applying Material

- Click on the **Apply Material** button from the **Ribbon**. The **Material** dialog box will be displayed as discussed in earlier chapters.
- Select the **Alloy Steel** option from the left list and click on the **Apply** button.
- Click on the **Close** button from the dialog box to exit. The color of model will be changed; refer to Figure-3.

Figure-3. Model after applying material

Applying Fixtures

- Click on the down arrow below **Fixtures Advisor** in the **Ribbon**. The tools related to fixtures will be displayed.
- Click on the **Fixed Geometry** button from the tool list. The **Fixture PropertyManager** will be displayed.
- Select the internal round faces of the model; refer to Figure-4.

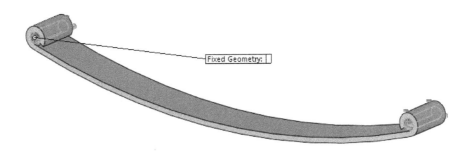

Figure-4. Faces to be fixed

- Click on the **OK** button from the **PropertyManager**. The round faces will be fixed as if they are bolted at that position.

Applying Force Vibrations

- Click on the down arrow below **External Loads Advisor** button in the **Ribbon**. List of the tools related to loads will be displayed.
- Click on the **Force** button from the list. The **Force/Torque PropertyManager** will be displayed.
- Click in the **Force Value** edit box in **Force/Torque** rollout and specify the value as **5000**.
- Select the face of the model as shown in Figure-5.

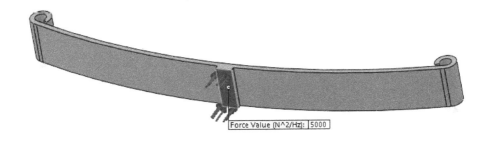

Figure-5. Face of model to be selected

- Click on the **OK** button from the **PropertyManager**.

Applying Global Damping

- Click on the **Global Damping** button in the **Ribbon**. The **Global Damping PropertyManager** will be displayed; refer to Figure-6.

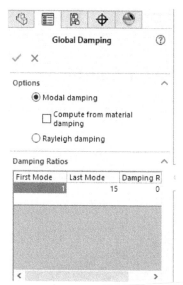

Figure-6. Global Damping PropertyManager

- Specify the modal damping ratio in the cell under **Damping Ratios** column in the table or click on the **Rayleigh damping** radio button and specify the desired parameters in the edit boxes displayed.
- Click on the **OK** button from the **PropertyManager**.

Running the analysis

- Click on the **Run** button from the **Ribbon**. The system will start analyzing the problem. Once the system is done, the results will be displayed in the **Modeling** area; refer to Figure-7.

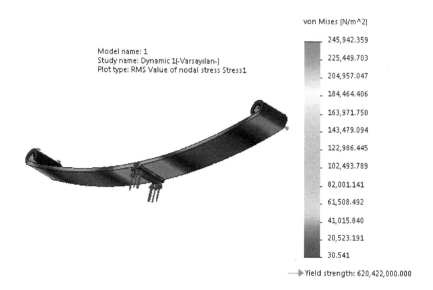

Figure-7. Result

- You may need to deselect the **Deformed Result** button from the **Ribbon** to check the non-deformed result; refer to Figure-8.

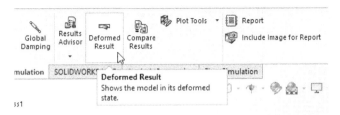

Figure-8. Deformed Result button

You can generated the result force, resonant frequencies, mass participation, Mode shape and other result plots as discussed in earlier chapters.

Properties for Random Vibration Study

After starting the study, click on the **Study Properties** tool from the **New Study** drop-down in the **Ribbon**. The **Random Vibration** dialog box will be displayed; refer to Figure-9.

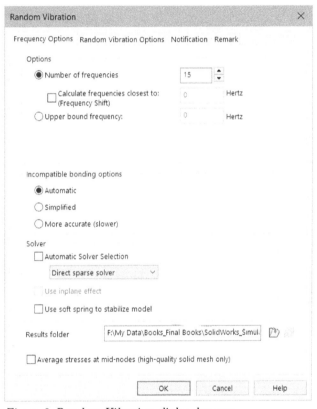

Figure-9. Random_Vibration_dialog_box.png

- Specify the desired number of natural frequencies in the **Number of frequencies** spinner. Note that increasing the compressive load on model will decrease the values of natural frequencies and if you increase the tensile load on the model then values of natural frequencies will increase.
- Select the **Calculate frequencies closest to** check box and specify the desired value of frequency near by which you want to find natural frequencies in next edit box.
- Set the other parameters as discussed earlier in **Frequency Options** tab. For most of the random vibration analyses, you need to use Direct Sparse solver with automatic incompatible bonding.

- Click on the **Random Vibration Options** tab to define parameters related to the analysis. The dialog box will be displayed as shown in Figure-10.

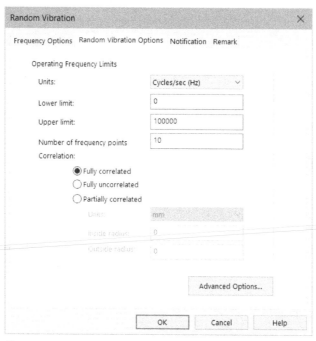

Figure-10. Random_Vibration_Options_tab.png

- Select the desired unit for frequency from the **Units** drop-down.
- Specify the desired upper and lower limits for the base frequencies (frequencies at which load is being applied).
- Specify the desired value in **Number of frequency points** edit box to define the number of frequencies at which loads are applied on the model.
- The radio buttons in the **Correlation** area are used to define how base frequency and load will be correlated in analysis. Select the desired radio button to define correlation.
- Click on the **Advanced Options** button to define advanced parameters for study. The dialog box will be displayed as shown in Figure-11.

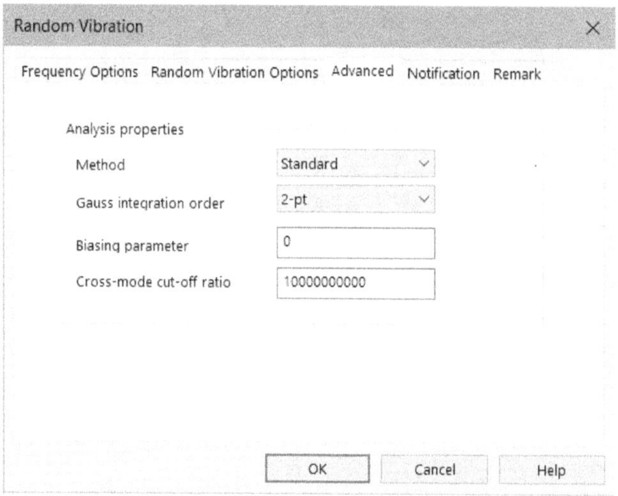

Figure-11. Advanced_tab.png

- Select the desired method of integration for base frequency and load from the **Method** drop-down. The **Standard** method is slower but accurate. The **Approximate** method is faster but gives relatively less accuracy. In **Approximate** method, the cross mode response are 0 and psd (power spectral density) of excitation are constant.
- Set the desired value of integration order from the **Gauss integration order** drop-down. The integration of loads with frequency points will be done at specified integration order.
- Specify the desired value in **Biasing parameter** edit box to define how frequency points will be selected in frequency range. If value of biasing parameter is 1 then frequency points will be uniformly distributed between natural frequencies. If biasing parameter is more than 1 then points will be closer to natural frequencies. If the values are less than 1 then frequency points will be far from natural frequencies.
- Modal psds of response are evaluated at each of the frequency points. The **cross-mode cut-off ratio** (RATIO) sets a limit on the ratio of all possible pairs of natural frequencies (w_i / w_j, $i > j$). This means that for each pair of modes with $w_i / w_j >$ RATIO, the cross-spectral density terms are neglected. Cross-mode effects are not considered for RATIO $=1$.

The other options in this dialog box are same as discussed earlier. Click on the **OK** button from the dialog box to apply parameters.

MODAL TIME HISTORY ANALYSIS

As discussed earlier, Modal Time History analysis is done when you know the variation of load with respect to time explicitly. Note that Time has good relation with Frequency which is Frequency = 1/Time. So, we can define the changing load with respect to time or Frequency. The procedure to perform this analysis is given next.

Assume that there is a bolt mounted on engine chamber. Now, you know that engine has little bit teasing nature when it is started. So, the force exerted on bolt is varying with time. The complete parameters of study are given in Figure-12. We need to find out the Stress induced in the bolt.

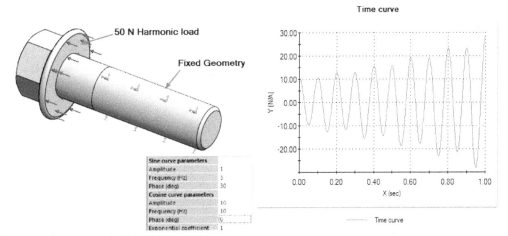

Figure-12. Parameters for Modal Time History study

Starting Modal Time History Analysis

- Open the part on which you want to perform the analysis.
- Click on the **SolidWorks Simulation** button from the **SOLIDWORKS Add-Ins** tab of the **Ribbon** to add **Simulation** tab in the **Ribbon**, if not added already.
- Click on the Down arrow below the **Study Advisor** button and select the **New Study** tool from the drop-down. List of analysis studies that can be performed, will be displayed in the left.
- Click on the **Linear Dynamic** button and then on the **Modal Time History** button from the **Options** rollout; refer to Figure-13.

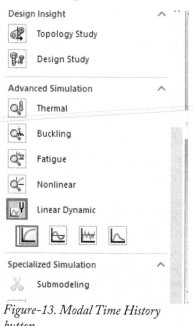

Figure-13. Modal Time History button

- Click on the **OK** button from the **PropertyManager**. The options related to Modal Time History Analysis will be displayed.

Applying Material and Fixture

- Click on the **Apply Material** button from the **Ribbon** and select the **Alloy Steel** material from the **Material** dialog box displayed.
- Click on the **Apply** button and then **Close** button from the **Material** dialog box.
- Click on the down arrow below **Fixtures Advisor** button in the **Ribbon** and select the **Fixed Geometry** button. The **Fixture PropertyManager** will be displayed.
- Select the round bottom face of the bolt as shown in Figure-14.

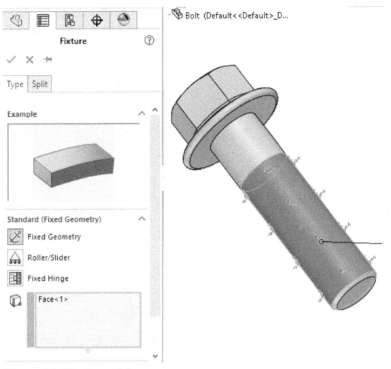

Figure-14. Fixed round face

- Click on the **OK** button from the **PropertyManager** to apply fixture.

Applying Harmonic Force

- Click on the down arrow below **External Loads Advisor** button in the **Ribbon**. List of tools will be displayed.
- Click on the **Force** button from the list. The **Force/Torque PropertyManager** will be displayed.
- Select the flat face of the bolt as shown in Figure-15.

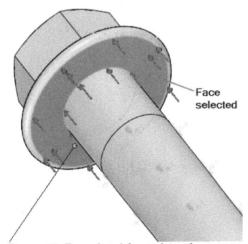

Figure-15. Face selected for applying force

- Click in the **Force Value** edit box in the **PropertyManager** and specify the value as **50** N.
- Select the **Curve** radio button from the **Variation with Time** rollout in the **PropertyManager** and click on the **Edit** button. The **Time Curve** dialog box will be displayed.

- Select **Harmonic Loading** option from the **Shape** drop-down and specify the parameters as shown in Figure-16. Specify the **Start time** and **End time** as **0** and **1** respectively, in the dialog box.

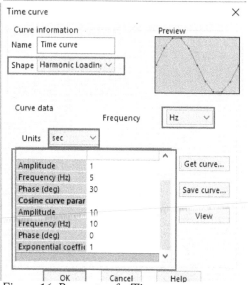

Figure-16. Parameters for Time curve

- Click on the **OK** button from the dialog box to apply the Harmonic curve.
- Click on the **OK** button from the **PropertyManager** to apply the force.

Running the Analysis

Click on the **Run This Study** button from the **Ribbon** to see the magic. Stress result will be displayed as shown in Figure-17. You already know the rest of story about generating other reports like Modal Shape Plot, Frequency Response Plot, and so on.

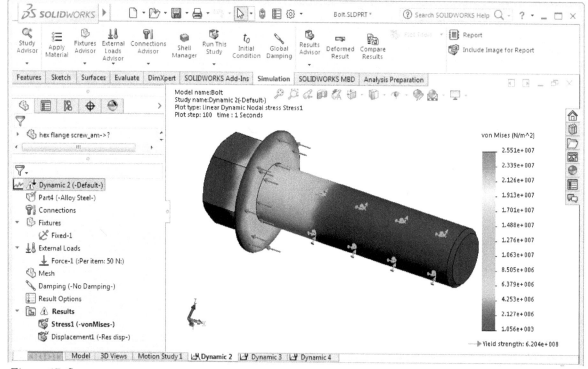

Figure-17. Stress report

In the above analysis, we have seen the effect of vibration on the bolt during starting of engine. We find that the stress induced is well below the Yield Strength. But, the journey of bolt has not ended here. Now, the bolt has to sustain the continuous harmonic loading due to vibrations of engine because Jimmy is going to New York riding his bike.

Modal Time History Properties

After starting Modal Time History study, click on the **Study Properties** tool from the **New Study** drop-down in the **Ribbon**. The **Modal Time History** dialog box will be displayed. The options in **Frequency Options** tab are same as discussed earlier. Click on the **Dynamic Options** tab in the dialog box. The options will be displayed as shown in Figure-18.

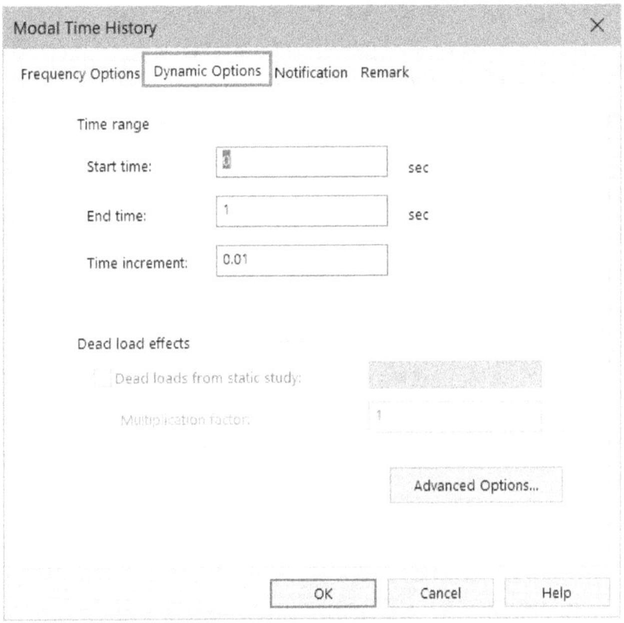

Figure-18. Dynamic_Options_tab.png

Specify the desired starting and end time in seconds for analysis in **Start time** and **End time** edit boxes. Specify the desired value of time step interval in the **Time increment** edit box. The calculations will be performed at each time step. Click on the **Advanced Options** button. The options of **Advanced** tab will be displayed; refer to Figure-19.

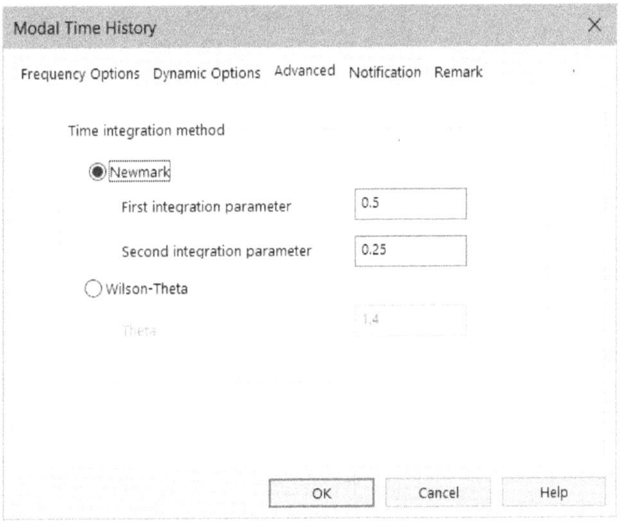

Figure-19. Advanced_options_for_Modal_Time_History.png

Select the desired option from the **Time integration method** area to define the method of integration for analysis. The integration methods have been discussed earlier.

HARMONIC ANALYSIS

Now, we have a range of frequency under which the bolt is going to work during his course to New York, say **0 Hz** to **5000 Hz**. Note that there might be natural frequencies of bolt in this range which can be dangerous for Jimmy's safety. Let's see whether Jimmy will reach New York or not.

Starting Harmonic Analysis

- Open the part on which you want to perform the analysis.
- Click on the **SolidWorks Simulation** button from the **SOLIDWORKS Add-Ins** tab of the **Ribbon** to add **Simulation** tab in the **Ribbon**, if not added already.
- Click on the Down arrow below the **Study Advisor** button and select the **New Study** tool from the drop-down. List of analysis studies that can be performed, will be displayed in the left.
- Click on the **Linear Dynamic** button and then on the **Harmonic** button from the **Options** rollout; refer to Figure-20.

Figure-20. Harmonic Study button

- Click on the **OK** button from the **PropertyManager**. The options related to Harmonic Analysis will be displayed.

Applying Material and Fixture

- Click on the **Apply Material** button from the **Ribbon** and select the **Alloy Steel** material from the **Material** dialog box displayed.
- Click on the **Apply** button and then **Close** button from the **Material** dialog box.
- Click on the down arrow below **Fixtures Advisor** button in the **Ribbon** and select the **Fixed Geometry** button. The **Fixture PropertyManager** will be displayed.
- Select the round bottom face of the bolt as shown in Figure-21.

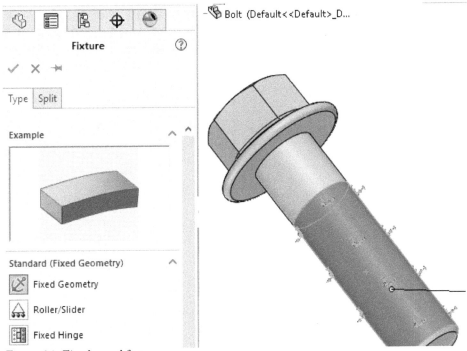

Figure-21. Fixed round face

- Click on the **OK** button from the **PropertyManager** to apply fixture.

Applying Force

- Click on the down arrow below **External Loads Advisor** button in the **Ribbon**. List of tools will be displayed.
- Click on the **Force** button from the list. The **Force/Torque PropertyManager** will be displayed.
- Select the flat face of the bolt as shown in Figure-22.

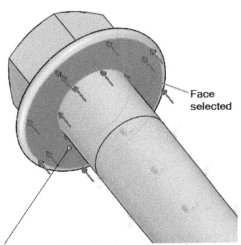

Face selected

Figure-22. Face selected for applying force

- Click in the **Force Value** edit box in the **PropertyManager** and specify the value as **50** N.
- Click in the **Phase Angle** edit box in **Phase Angle** rollout if you want to specify phase angle. (We have kept it zero. Remember the formula; $P = A \sin (\omega t + \varphi)$ where φ is phase angle).
- Click on the **OK** button from the **PropertyManager** to apply the force.

Setting Frequency Range

- Right-click on the Harmonic study from the **Analysis Manager** and select the **Properties** option from the shortcut menu; refer to Figure-23. The **Harmonic** dialog box will be displayed; refer to Figure-24.

Figure-23. Properties option

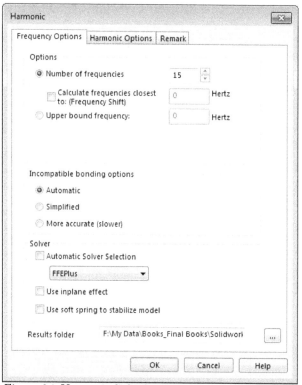

Figure-24. Harmonic dialog box

- Specify the number of frequencies to be tested to **10** by using the **Number of frequencies** spinner in the **Options** area of the dialog box.
- Click on the **Harmonic options** tab and specify the **Upper limit** as 5000 Hz; refer to Figure-25.

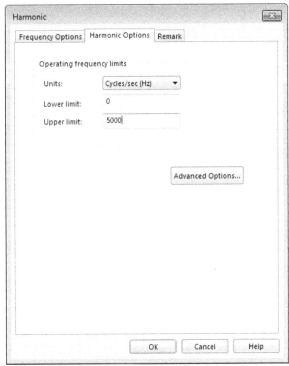

Figure-25. Upper limit for frequencies

- Click on the **OK** button from the dialog box.

Jimmy reaches or not (Running Analysis)

- Click on the **Run This Study** button from the **Ribbon** and wait & watch. The Stress result will be displayed as shown in Figure-26 which concludes that nobody can stop Jimmy. The stress in bolt is well below the yield strength.

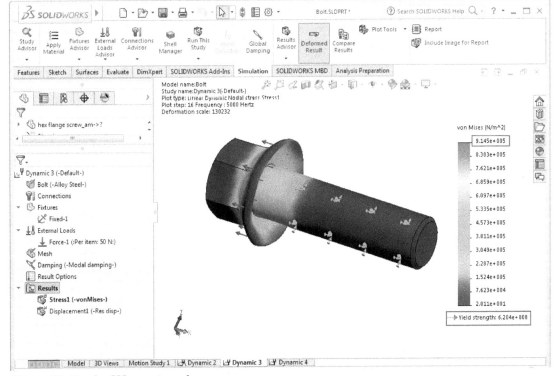

Figure-26. Result of Harmonic study

Harmonic Analysis Properties

After starting the harmonic analysis, click on the **Study Properties** tool from the **New Study** drop-down in the **Ribbon**. The **Harmonic** dialog box will be displayed to define harmonic properties. The options in the **Frequency Options** tab are same as discussed earlier. Click on the **Harmonic Options** tab to define parameters related to frequency limits; refer to Figure-27. Specify the lower and upper limits for frequencies to be tested for natural frequency. Click on the **Advanced Options** button from the dialog box. The options of **Advanced** tab will be displayed; refer to Figure-28.

Figure-27. Harmonic_Options_tab.png

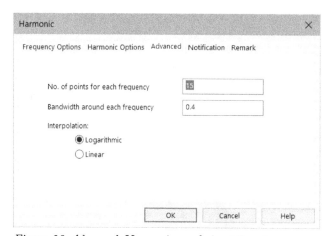

Figure-28. Advanced_Harmonic_analysis_options.png

- Specify the desired number of frequency points at which responses are to be calculated in the **No. of points for each frequency** edit box.
- Specify the desired bandwidth ratio within which you want to consider frequencies for performing analysis in the **Bandwidth around each frequency** edit box.
- Select the desired radio button from the Interpolation area to define how frequency points will be selected in the range.

Note

A solution frequency point is used at each natural frequency considered in the analysis. If two consecutive modes are too close to each other, the software uses a number of points smaller than No. of points for each frequency. For the first mode, F_{i-1} is considered 0. For the last mode, the bandwidth is considered to be symmetrical around the natural frequency. If the upper limit frequency, specified in the **Harmonic Options** tab, is much larger than the highest natural frequency considered in the analysis, the No. of points for each frequency input is used to define frequency points in that range.

There are some situations where you need to see the effect of base excitation on the object. In such cases, you should perform the Response Spectrum analysis. The process is given next.

RESPONSE SPECTRUM ANALYSIS

Let me tell you that Jimmy placed his camera in the side box of his motorbike by mistake and the camera has experienced all the jerks provided by road. Let's see what happens to the body of his new camera. Assume that the excitation caused on camera body is 2 m/s^2.

Starting Response Spectrum Analysis

- Open the part on which you want to perform the analysis (obviously camera body!!).
- Click on the down arrow below the **Study Advisor** button and select the **New Study** tool from the drop-down. List of analysis studies that can be performed, will be displayed in the left.
- Click on the **Linear Dynamic** button and then on the **Response Spectrum Analysis** button from the **Options** rollout; refer to Figure-29.

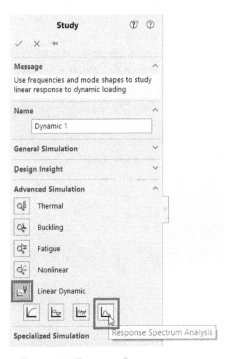

Figure-29. Response Spectrum Analysis button

- Click on the **OK** button from the **PropertyManager**. The options related to Response Spectrum Analysis will be displayed.

Applying Material and Fixture

- Click on the **Apply Material** button from the **Ribbon** and select the **Acrylic (Medium-high impact)** plastic from the **Material** dialog box displayed; refer to Figure-30.

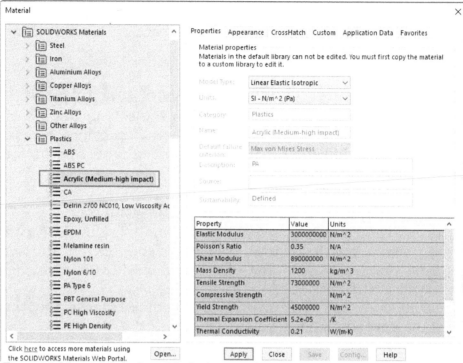

Figure-30. Material dialog box with plastic materials

- Click on the **Apply** button and then **Close** button from the **Material** dialog box.
- Click on the down arrow below **Fixtures Advisor** button in the **Ribbon** and select the **Fixed Geometry** button. The **Fixture PropertyManager** will be displayed.
- Select the bottom face of the camera as shown in Figure-31.

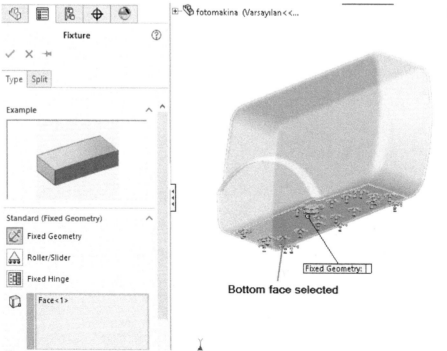

Figure-31. Bottom face of camera body selected

- Click on the **OK** button from the **PropertyManager** to apply fixture.

Applying Excitation

No sir! Excitation is not linked with human emotions here. Excitation is a displacement, velocity, or acceleration caused in the body due to vibrations. There are two tools available to define excitation; refer to Figure-32.

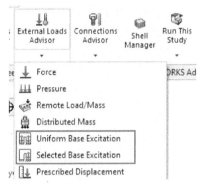

Figure-32. Excitation tools

If you have specified base at more than one faces then you can use the **Selected Base Excitation** tool to specify excitation for individual bases. To specify the same excitation for all the bases, use the **Uniform Base Excitation** tool.

- Click on the down arrow below **External Loads Advisor** button in the **Ribbon**. List of tools will be displayed.
- Click on the **Uniform Base Excitation** tool from the list. The **Uniform Base Excitation PropertyManager** will be displayed.
- Select the **Acceleration** radio button from the **Type** rollout and specify **2** in **Along Plane Dir 2** vertical direction; refer to Figure-33.

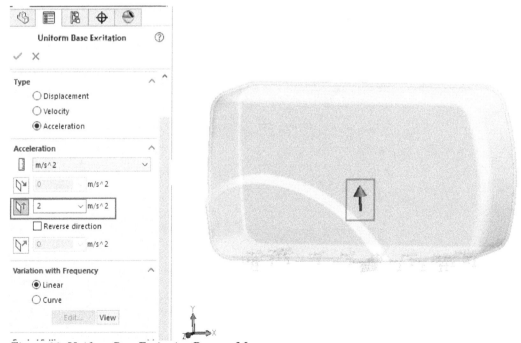

Figure-33. Uniform Base Excitation PropertyManager

- Click on the **OK** button from the **PropertyManager** to apply the excitation.

Now we are ready for the magic. Click on the **Run This Study** button from the **Ribbon** and see whether the camera is in one piece or more pieces; refer to Figure-34.

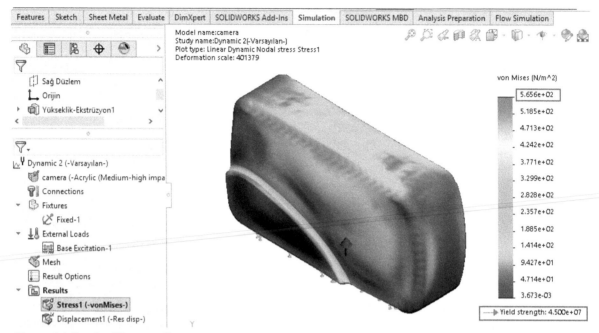

Figure-34. Result of Response Spectrum

Even if you change the value of Base excitation to **20000** m/s², still the camera will be safe. Because the power of acrylic plastic will not let you down. This plastic is used to make bullet proof shields.

Properties of Response Spectrum Analysis

After starting response spectrum analysis, click on the **Study Properties** tool from the **New Study** drop-down. The **Response Spectrum Analysis** dialog box will be displayed; refer to Figure-35.

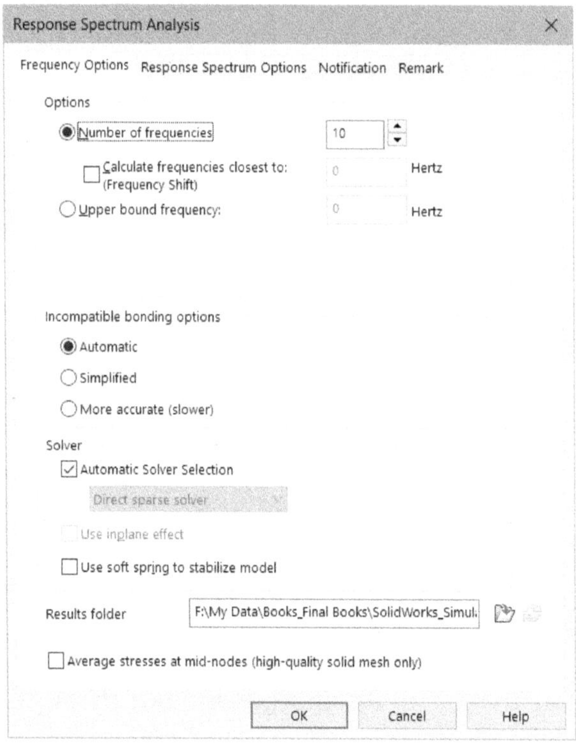

Figure-35. Response_Spectrum_Analysis_dialog_box.png

Specify the desired parameters in the **Frequency Options** tab as discussed earlier. Click on the **Response Spectrum Options** tab in the dialog box. The options will be displayed as shown in Figure-36.

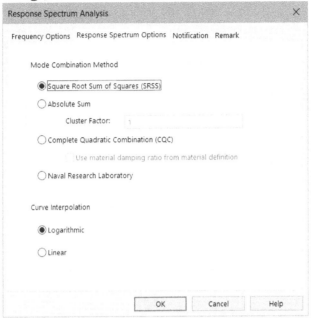

Figure-36. Response_Spectrum_Options_tab.png

Select the desired radio button from the **Mode Combination Method** area. These methods are discussed next.

Square Root Sum of Squares (SRSS)

In this method, square root of sum of maximum modal responses is combined to get peak response. Mathematically, it can be given as:

$$\{u\}_{max} = \sqrt{\sum_{i=1}^{nmodes} |\{u\}^i_{max}|^2}$$

Absolute Sum

This method assumes that the maximum modal responses occur at the same time for all modes. It is the most conservative among the modal combination methods.

$$\{u\}_{max} = \sum_{i=1}^{nmodes} |\{u\}^i_{max}|$$

Complete Quadratic Combination

This method is based on random vibration theories. The peak response is estimated from the maximum modal values from the double summation equation.

$$\{u\}_{max} = \sqrt{\sum_{i=1}^{nmodes} \sum_{j}^{nmodes} \{u\}^i_{max} * \{u\}^j_{max} * \rho_{ij}}$$

here, ρ_{ij} is the cross-modal correlation coefficient and ξ_i and ξ_j are modal damping coefficients for modes i and j.

$$\rho_{ij} = \frac{8\sqrt{\xi_i \xi_j}\,(\xi_i + r\xi_j)r^{3/2}}{(1-r^2)^2 + 4\xi_i \xi_j r(1+r^2) + 4(\xi_i^{\,2} + \xi_j^{\,2})r^2}$$

$$r = \frac{\omega_i}{\omega_j}$$

Naval Research Laboratory (NRL)

The mode combination method recommended by the NRL takes the absolute value of the response for the mode that exhibits the largest response and adds it to the SRSS response of the remaining modes. Mathematically,

$$\{u\}_{max} = \left|\{u_j\}_{max}\right| + \sqrt{\sum_{i=1}^{nmodes} [\{u\}_{max}^i]^2 - \{u_j\}_{max}^2}$$

here, $\{u_j\}_{max}$ represents the mode with the largest response among all modal responses.

PRACTICE 1

In the below problem, you will analyze a driving shaft of automobile for its ability to sustain the torque provided by engine. Note that the driver of this vehicle is rash driver and he really don't care about his machine. So, you might need to perform the random vibration analysis. The parameters for the analysis are given in Figure-37. The model of this problem is available in resource kit.

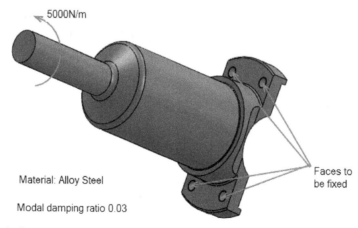

5000N/m

Material: Alloy Steel

Modal damping ratio 0.03

Faces to be fixed

Figure-37. Parameters for analysis

SELF-ASSESSMENT

Q1. Static studies assume that loads are constant or applied at an instant. (T/F)

Q2. Generally if the frequency of a load is larger than 1/3 of the lowest (fundamental) frequency of object, a dynamic study should be used. (T/F)

Q3. A viscous damper (or dashpot) generates a force that is proportional to mass. (T/F)

Q4. Critical damping C_c is the least amount of damping that causes a system to return to its equilibrium position without oscillating. (T/F)

Q5. If the part has softer elements then Rayleigh Damping can product unrealistic results because of stiffness matrix part. (T/F)

Q6. Which of the following dynamic analysis should be used to find out stress when variation of each load with time is known explicitly?

a. Modal Time History Analysis
b. Harmonic Analysis
c. Random Vibration Analysis
d. Response Spectrum Analysis

Q7. Which of the following dynamic analysis should be used to calculate the peak steady state response due to harmonic loads or base excitations.

a. Modal Time History Analysis
b. Harmonic Analysis
c. Random Vibration Analysis
d. Response Spectrum Analysis

Q8. Which of the following dynamic analysis should be used to calculate the response due to non-deterministic loads.
a. Modal Time History Analysis
b. Harmonic Analysis
c. Random Vibration Analysis
d. Response Spectrum Analysis

Q9. In a random vibration study, loads are described statistically by .... functions.

Q10. A displacement, velocity, or acceleration caused in the body due to vibrations is called ....

FOR STUDENT NOTES

Answer to Self-Assessment:
1. T, 2. T, 3. F, 4. T, 5. T, 6. a, 7. b, 8. c, 9. power spectral density (psd), 10. excitation

Chapter 9

Thermal Analysis

Topics Covered

The major topics covered in this chapter are:

- *Introduction*
- *Types of Thermal Analysis*
- *Starting Thermal Analysis.*
- *Applying Material.*
- *Defining Fixtures*
- *Applying loads*
- *Defining Connections*
- *Simulating Analysis.*
- *Interpreting results*

INTRODUCTION

Thermal analysis is a method to check the distribution of heat over a body due to applied thermal loads. Note that thermal energy is dynamic in nature and is always flowing through various mediums. There are three mechanisms by which the thermal energy flows:

- Conduction
- Convection
- Radiation

In all three mechanisms, heat energy flows from the medium with higher temperature to the medium with lower temperature. Heat transfer by conduction and convection requires the presence of an intervening medium while heat transfer by radiation does not.

The output from a thermal analysis can be given by:
1. Temperature distribution.
2. Amount of heat loss or gain.
3. Thermal gradients.
4. Thermal fluxes.

This analysis is used in many engineering industries such as automobile, piping, electronic, power generation, and so on.

Important terms related to Thermal Analysis

Before conducting thermal analysis, you should be familiar with the basic concepts and terminologies of thermal analysis. Following are some of the important terms used in thermal analysis:

Heat Transfer Modes

Whenever there is a difference in temperature between two bodies, the heat is transferred from one body to another. Basically, heat is transferred in three ways: Conduction,Convection,and Radiation.

Conduction

In conduction, the heat is transferred by interactions of atoms or molecules of the material.
For example, if you heat up a metal rod at one end, the heat will be transferred to the other end by the atoms or molecules of the metal rod.

Convection

In convection, the heat is transferred by the flowing fluid. The fluid can be gas or liquid. Heating up water using an electric water heater is a good example of heat convection. In this case, water takes heat from the heater.

Radiation

In radiation, the heat is transferred in space without any matter. Radiation is the only heat transfer method that takes place in space. Heat coming from the Sun is a good example of radiation. The heat from the Sun is transferred to the earth through radiation.

Thermal Gradient

The thermal gradient is the rate of increase in temperature per unit depth in a material.

Thermal Flux

The Thermal flux is defined as the rate of heat transfer per unit cross-sectional area. It is denoted by q.

Bulk Temperature

It is the temperature of a fluid flowing outside the material. It is denoted by Tb. The Bulk temperature is used in convective heat transfer.

Film Coefficient

It is a measure of the heat transfer through an air film.

Emissivity

The emissivity of a material is the ratio of energy radiated by the material to the energy radiated by a black body at the same temperature. Emissivity is the measure of a material's ability to absorb and radiate heat. It is denoted by e. Emissivity is a numerical value without any unit. For a perfect black body, e = 1. For any other material, e < 1.

Stefan–Boltzmann Constant

The energy radiated by a black body per unit area per unit time divided by the fourth power of the body's temperature is known as the Stefan-Boltzmann constant. It is denoted by s.

Thermal Conductivity

The thermal conductivity is the property of a material that indicates its ability to conduct heat. It is denoted by K.

Specific Heat

The specific heat is the amount of heat required per unit mass to raise the temperature of the body by one degree Celsius. It is denoted by c.

THERMAL LOADS

To start the thermal analysis, click on the down arrow below **New Study** button from **Ribbon** and select the **New Study** button from the tools displayed. The **Study PropertyManager** will be displayed. Click on the **Thermal** button from the **Type** rollout and click on the **OK** button from the **PropertyManager**. The tools related to thermal analysis will be displayed. Before performing thermal analysis, we will get familiar with various options required specifically for thermal analysis. One of the major player in thermal analysis is the thermal load. To apply the thermal loads, options are available in the **Thermal Loads** drop-down; refer to Figure-1. The options in this drop-down are discussed next.

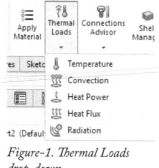

Figure-1. Thermal Loads drop-down

Temperature

The **Temperature** tool is used to specify temperature of selected face/faces. To apply temperature follow the steps given next.

- Click on the **Temperature** tool from the list of tools displayed on click down arrow below **Thermal Loads** in the **Ribbon**. The **Temperature PropertyManager** will be displayed as shown in Figure-2.

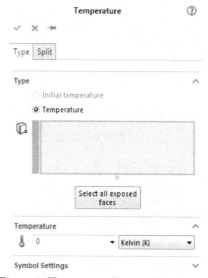

Figure-2. Temperature PropertyManager

- Select the face on which you want to set the temperature.
- Click in the **Temperature** edit box and specify the desired value.

- If you want to change the unit of temperature from the default Kelvin unit, then click on the **Kelvin (K)** option next to **Temperature** edit box. The list of options related to temperature units will be displayed.
- Select the desired unit from the list.
- If you are performing the transient thermal study, then the **Initial temperature** radio button will be active; refer to Figure-3.

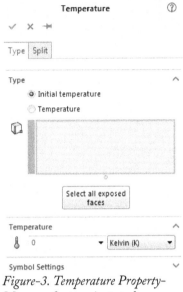

Figure-3. Temperature Property-Manager for transient study

How to start Transient Thermal Analysis? After starting the thermal analysis, right-click on the analysis name in the **Analysis Manager**; refer to Figure-4. The options related to analysis will be displayed in the shortcut menu. Select the **Properties** option from the menu. The **Thermal** dialog box will be displayed; refer to Figure-5. Select the **Transient** radio button from the **Solution Type** area of the dialog box. The options related to analysis time will be displayed. Set the desired time duration and increment for the analysis by using the edit boxes. The other options in this dialog box have already been discussed in previous chapters. Click on the **OK** button from the dialog box to apply the settings.

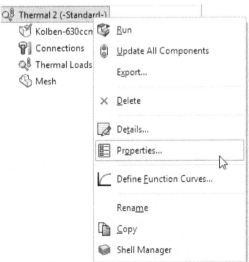

Figure-4. Analysis Properties shortcut menu

Figure-5. *Thermal dialog box*

- Using this option, you can specify the starting temperature of the face. Although, it will change automatically under the application of heat flux. You will learn more about this tool later while performing the analysis.

- If you select the **Temperature** radio button for a transient thermal study then the **Variation with Time** rollout is also displayed in the **PropertyManager**; refer to Figure-6.

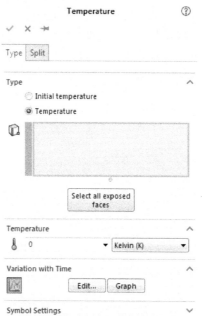

Figure-6. *Temperature PropertyManager with time varying temperature*

- The options in the **Variation with Time** rollout are same as discussed in previous chapters.
- If you want to select the all exposed faces of the model, then click on the **Select all exposed** faces button from the **PropertyManager**.
- After specifying the desired parameters, click on the **OK** button from the dialog box.

Convection

Convection is the phenomena by which heat is transferred through a medium like air, fluid or solid. Every medium has its limitation to transfer the heat energy. This limitation is mathematically represented by **Convection Heat coefficient** $Q_{convection}$.

$$Q_{convection} = hA(T_s-T_f)$$

Here, **h** is heat transfer coefficient has unit $W/m^2.k$
    T_s is the temperature of the surface.
    T_f is the temperature of the surrounding fluid.
    A is the area of surface.

In SolidWorks Simulation, we need to specify the $Q_{convection}$ for the convection phenomena. The steps to apply the heat convection are given next.

- Click on the down arrow below the **Thermal Loads** from the **Ribbon**. The list of tools will be displayed.
- Select the **Convection** button from the list of tools. The **Convection PropertyManager** will be displayed as shown in Figure-7.

Figure-7. Convection PropertyManager

- Select the face from which the convection is occurring.
- Click in the **Convection Coefficient** edit box in the **Convection Coefficient** rollout and specify the desired value in it.

- Click in the **Bulk Ambient Temperature** edit box in the **Bulk Ambient Temperature** rollout and specify the surrounding temperature for the analysis.
- After specifying the desired parameters, click on the **OK** button from the **PropertyManager** to apply the heat convection.

Heat Power

This is the energy applied on the selected face in the form of heat. You can specify the heat power applied on the face by using the **Heat Power** button. The steps to use this button are given next.

- Click on the down arrow below the **Thermal Loads** button in the **Ribbon**. The list of tools will be displayed.
- Click on the **Heat Power** button from the list. The **Heat Power PropertyManager** will be displayed; refer to Figure-8.
- Select the face on which you want to apply the heat.
- Click in the **Heat Power** edit box in the **Heat Power** rollout in **PropertyManager** and specify the desired value of heat power.

Figure-8. Heat Power Property-Manager

- After specifying the desired parameters, click on the **OK** button from the **PropertyManager**. Note that for Transient study there will be some extra options in this **PropertyManager** like Temperature curve button and Thermostat (Transient) options.

Heat Flux

Heat flux is the heat energy applied per unit area of the selected face. The steps to apply heat flux on the model are given next.

- Click on the down arrow below **Thermal Loads** button in the **Ribbon**. The list of tools will be displayed.
- Click on the **Heat Flux** button from the list. The **Heat Flux PropertyManager** will be displayed as shown in Figure-9.

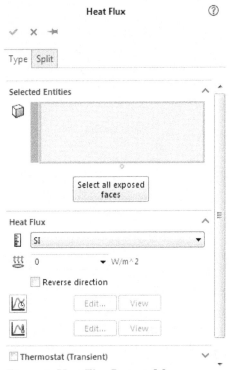

Figure-9. Heat Flux PropertyManager

- The options in this **PropertyManager** are same as in **Heat Power PropertyManager** and can be used in the same way.

Radiation

Thermal radiation is the thermal energy emitted by bodies in the form of electromagnetic waves because of their temperature. All bodies with temperatures above the absolute zero emit thermal energy. Because electromagnetic waves travel in vacuum, no medium is necessary for radiation to take place. Thermal radiation can occur between **ambient temperature surrounding and surface of body** or between **surface of one body and surface of other body**. The steps to specify radiation are given next.

- Click on down arrow below **Thermal Loads** button. The list of tools will be displayed.
- Click on the **Radiation** button from the list. The **Radiation PropertyManager** will be displayed as shown in Figure-10.

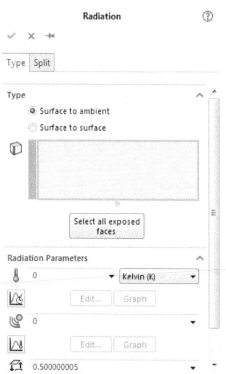

Figure-10. Radiation PropertyManager

- Select the face from which the radiation is occurring.
- Click in the first edit box in the **Radiation Parameters** rollout and specify the ambient temperature if you have selected the **Surface to ambient** radio button from the **Type** rollout.
- If you have selected the **Surface to surface** radio button from the **Type** rollout and selected the **Open system** check box from the **Radiation Parameters** rollout then also you can specify the ambient temperature in the first edit box in the **Radiation Parameters** rollout.
- Click in the **Emissivity** edit box and specify the emission coefficient for radiation.
- After specifying the desired parameters, click on the **OK** button from the **PropertyManager**.

CONTACT SET

We have used connections between various parts of assembly while performing structural analyses. In the same way, we are also required to specify connections between various parts of assembly in terms of thermal properties. For thermal analyses, we can specify the following connections:

- Thermal Resistance
- Bonded
- Insulated

The tools to specify these connections are available in the **Connections Advisor** drop-down. Click on the **Contact Set** tool from the drop-down. The **Contact Sets PropertyManager** will be displayed as shown in Figure-11.

Select the desired contact type from the drop-down in the **Type** rollout and select the faces to apply the thermal contact type. In the same way, you can use the **Component Contact** button to apply the connection between two parts.

Thermal Resistance Contact

The thermal resistance contact is used to specify thermal resistance between two components connected in assembly. On selecting this option, the **PropertyManager** is displayed as shown in Figure-11. Select the two contacting faces of the components in assembly by using the selection boxes in the **PropertyManager**. Specify the desired value of thermal resistance. The unit of thermal resistance is K/W.

Bonded Contact

The bonded contact is used when there is perfect heat conduction between two contacting parts. Note that the heat flow between parts will depend on their thermal conductance.

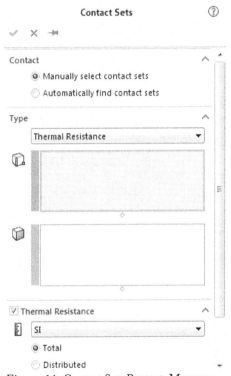

Figure-11. Contact Sets PropertyManager

Insulated Contact

The insulated contact is same as No penetration in structural analysis. This contact is used when there is no heat conduction between the two contacting parts.

PRACTICAL ON STEADY STATE THERMAL ANALYSIS

In this tutorial, we will perform a steady state thermal analysis on a real world model and check the distribution of temperature/heat over the surface of the model. Note that model for this practical is available in the resource kit.

Starting the analysis

- Open the part on which you want to perform the analysis.
- Click on the **SolidWorks Simulation** button from the **SOLIDWORKS Add-Ins** tab of the **Ribbon** to add **Simulation** tab in the **Ribbon**, if not added already.
- Click on the Down arrow below the **Study Advisor** button and select the **New Study** tool from the drop-down. List of analysis studies that can be performed, will be displayed in the left.
- Click on the **Thermal** button then click on the **OK** button from the **PropertyManager**. The tools related to thermal analysis will be displayed.

Applying Material

- Click on the **Apply Material** button from the **Ribbon**. The **Material** dialog box will be displayed.
- Expand the **SolidWorks Materials** node and then the **Aluminium Alloys** node in the left of the dialog box.
- Browse through the materials and select the **C355.0-T61 Permanent Mold cast (SS)** material; refer to Figure-12.
- Click on the **Apply** button and then click on the **Close** button from the dialog box to exit the dialog box.

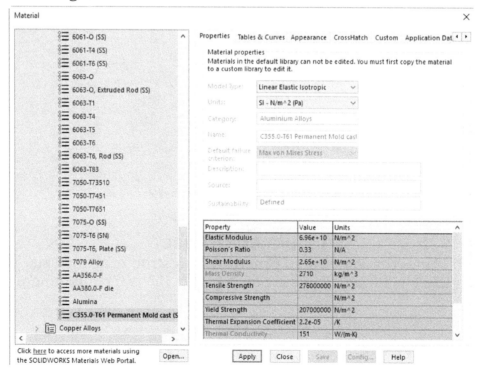

Figure-12. Material dialog box

Applying convection to the model

- Click on the down arrow below **Thermal Loads** button and select the **Convection** button from the list. The **Temperature PropertyManager** will be displayed as shown in Figure-13.

Figure-13. Convection PropertyManager

- Select all the inner faces of the model as shown in Figure-14.

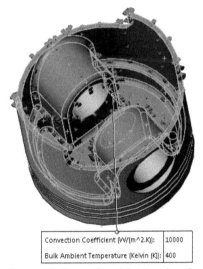

Figure-14. Faces of model to be selected

- Specify the convection coefficient as 10000 in the **Convection Coefficient** edit box in **PropertyManager**. This convection coefficient is for engine oil.
- Click in **Bulk Ambient Temperature** edit box in the **PropertyManager** and specify the value as 400 K.
- Click on the **OK** button from the **PropertyManager** to apply the convection.

Applying heat generated by gas combustion on the head of the piston

- Click on the down arrow of **Thermal Loads** in the **Ribbon** and select the **Heat Power** button from the list displayed. The **Heat Power PropertyManager** will be displayed; refer to Figure-15.

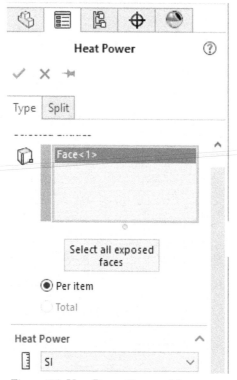

Figure-15. Heat Power PropertyManager

- Click on the **Total** radio button in the **Selected Entities** rollout.
- Select the head faces of the piston as shown in Figure-16.

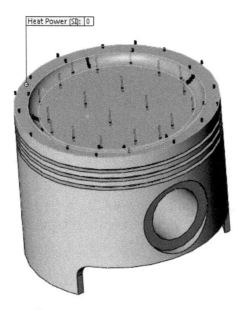

Figure-16. Faces to be selected

- Click in the **Heat Power** edit box in the **Heat Power** rollout. Specify the total heat power as **20000**.
- Click on the **OK** button from the **PropertyManager** to apply the heat.

Running the Analysis

- Click on the **Run** button from the **Ribbon**. The analyzing of problem will start.
- After the analysis is complete. The solution of the analysis will be displayed;refer to Figure-17.

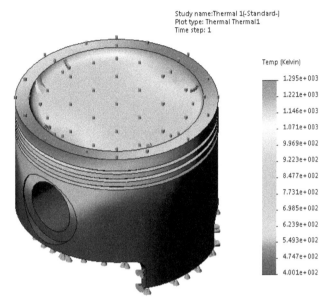

Study name:Thermal 1(-Standard-)
Plot type: Thermal Thermal1
Time step: 1

Temp (Kelvin)

1.295e+003
1.221e+003
1.146e+003
1.071e+003
9.969e+002
9.223e+002
8.477e+002
7.731e+002
6.985e+002
6.239e+002
5.493e+002
4.747e+002
4.001e+002

Figure-17. Solution of analysis

- To change the temperature scale, right-click on the Temperature scale at the right. A shortcut menu will be displayed; refer to Figure-18.

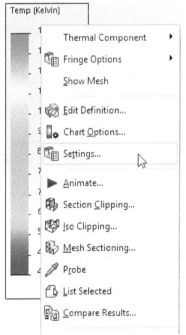

*Figure-18. Shortcut menu for tem-
perature scale*

- Click on the **Settings** button from the shortcut menu. The **Thermal Plot PropertyManager** will be displayed; refer to Figure-19.

- Click on the **Definition** tab and then click on the second drop-down in the **Display** rollout; refer to Figure-20.

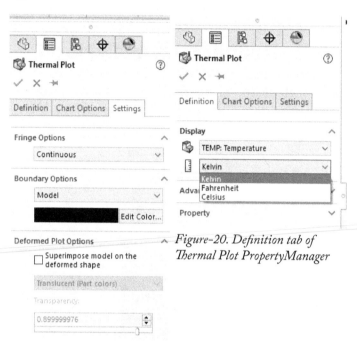

Figure-19. Thermal Plot
PropertyManager

Figure-20. Definition tab of
Thermal Plot PropertyManager

- Click on the **Celsius** option from the list displayed and click on the **OK** button from the **PropertyManager**. The scale will change to **Celsius**.
- You can select the other components of Thermal analysis by right-clicking on the **Temperature scale** and selecting the desired option from the Thermal Component cascading menu; refer to Figure-21.

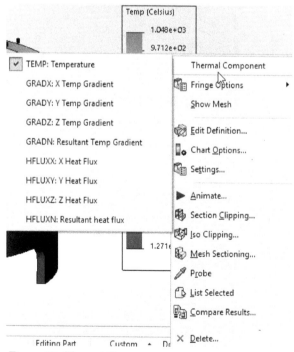

Figure-21. Thermal Components

SWITCHING FROM STEADY STATE ANALYSIS TO TRANSIENT THERMAL ANALYSIS

Earlier, we have performed the steady state analysis. Now, we will use the above analysis and switch to transient thermal analysis. The procedure to switch to transient thermal analysis is given next.

- Right-click on the name of analysis in the **Analysis Manager**; refer to Figure-22.
- Click on the **Properties** button from the menu displayed. The **Thermal** dialog box will be displayed; refer to Figure-23.

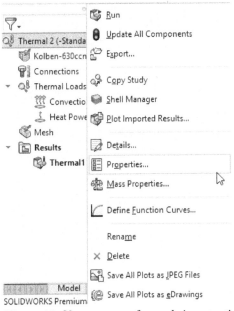

Figure-22. Shortcut menu for analysis properties

Figure-23. Thermal dialog box

- Click on the **Transient** radio button. The edit boxes below it will become active.
- Specify the total time of study (in seconds) in the **Total time** edit box as 10 seconds.
- Click in the **Time increment** edit box and specify the increment value in seconds as 1.
- If you have earlier performed any other analysis from where you want to import the temperature as the initial value then click on the **Initial temperature from thermal study** check box and then select the study name from the drop-down.
- You can also select the desired step number from the adjacent spinner.
- Select the desired solver from the **Solver** area of the dialog box and then click on the **OK** button from the dialog box.

Specifying Initial Temperature

Now, we need to specify the initial temperature for the analysis if we have not imported it from any earlier study.

- Click on the **Temperature** button from the **Thermal Loads** drop-down. The **Temperature PropertyManager** will be displayed.
- Select the **Initial Temperature** radio button and select all the faces by using Select all exposed faces button as shown in Figure-24.

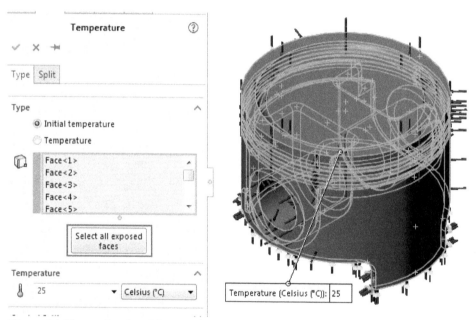

Figure-24. Faces to be selected for initial temperature

- Click in the **Temperature** edit box and in the **Temperature** rollout and specify the value as **25 °C**.
- Click on the **OK** button from the **PropertyManager** to apply the temperature.

Running the analysis

- Click on the **Run This Study** button to run the analysis.
- After the analysis is complete, compare the two results. You will find that the analysis temperature range is changing and the part is displayed cooler than earlier one. This is because we have allowed the time for cooling in analysis and we have specified the initial temperature as **25°C**.

- When you animate the result, we will find that heat is transferred from top face to side walls as the time passes.

Using Probe to find out temperature

- After running the analysis, right-click in the modeling area. A shortcut menu will be displayed.
- Select the **Probe** button from the shortcut menu; refer to Figure-25.

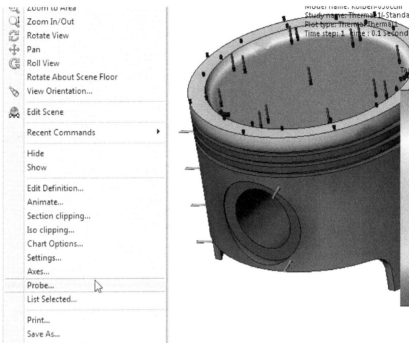

Figure-25. Probe button

- The **Probe Result PropertyManager** will be displayed; refer to Figure-26.

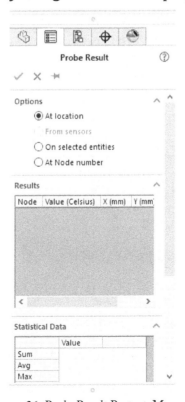

Figure-26. Probe Result PropertyManager

- Click on the model at the desired position. The temperature and other parameters of the selected position will be displayed in a box and **Results** rollout of **PropertyManager**; refer to Figure-27.

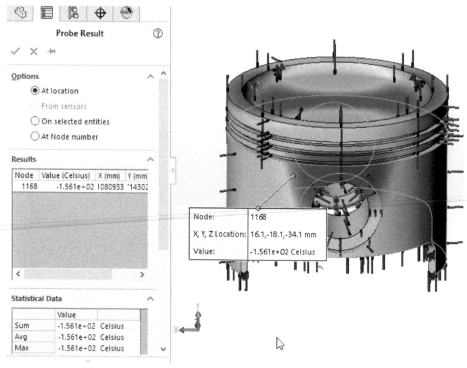

Figure-27. Temperature output

- Click on the **Plot** button from the **Report Options** rollout of the **PropertyManager** to check the graph of temperature change.

Note that to be able to perform the analysis using the SolidWorks Simulation software, you should be able to answer the following questions:

1. **Heat transfer through a solid body is referred to as**

a. Conduction b. Convection c. Radiation d. Generation

2. **The temperature gradient is defined as**

a. Temperature rate of change per unit length
b. Heat flow through a body
c. Temperature rate of change per unit volume
d. Temperature rate of change per unit area

3. **Heat flux is defined by**

a. The amount of heat generated per unit volume
b. The heat transfer rate per unit area
c. The amount of heat stored in a control volume
d. The temperature change per unit length

4.   Heat transfer between a solid body and a fluid is referred to as

a.   Conduction b. Convection c. Radiation d. Generation

5.   Which behavior best describes a material with a high thermal conductivity compared to a material will a smaller thermal conductivity?

a.   Temperature change is smaller through the solid
b.   Heat flux is higher through the solid
c.   Thermal resistance is lower through the solid
d.   All of the above

6.   The convection coefficient is related to

a.   The rate that heat is transferred between a solid and fluid
b.   The rate that heat is transferred between two solids
c.   The emissive power of the material
d.   The heat generation capability of the material

7.   The amount of heat transferred by radiation is directly related to

a.   The Stephan-Boltzmann constant
b.   The surface temperature
c.   The emissivity of the material
d.   All of the above

8.   If no heat is transferred to or from a surface, it is referred to as

a.   Isothermal b. Exothermal c. Adiabatic d. Isobaric

9.   When all temperatures are in equilibrium, the problem is assumed to be

a.   Transient
b.   Steady-State
c.   Laminar
d.   None of the above

10.   Which of the following terms is not part of the energy balance for heat transfer?

a.   Generated energy
b.   Stored energy
c.   Energy into the system
d.   Energy destroyed

****

PROBLEM 1

A metal sphere of diameter d = 35mm is initially at temperature Ti = 700 K. At t=0, the sphere is placed in a fluid environment that has properties of T∞ = 300 K and h = 50 W/m2-K. The properties of the steel are k = 35 W/m-K, ρ = 7500 kg/m3, and c = 550 J/kg-K. Find the surface temperature of the sphere after 500 seconds.

PROBLEM 2

A flanged pipe assembly; refer to Figure-28, made of plain carbon steel is subjected to both convective and conductive boundary conditions. Fluid inside the pipe is at a temperature of 130°C and has a convection coefficient of hi = 160 W/m²-K. Air on the outside of the pipe is at 20°C and has a convection coefficient of ho = 70 W/m²-K. The right and left ends of the pipe are at temperatures of 450°C and 80°C, respectively. There is a thermal resistance between the two flanges of 0.002 K-m²/W. Use thermal analysis to analyze the pipe under both steady state and transient conditions.

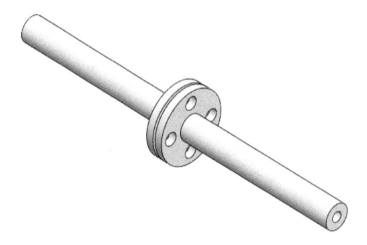

Figure-28. Flanged pipe assembly

FOR STUDENT NOTES

FOR STUDENT NOTES

Answers to Self-Assessment:
1. a, 2. a, 3. b, 4. b, 5. d, 6. a, 7. d, 8. c, 9. b, 10. d

Chapter 10

Buckling Analysis

Topics Covered

The major topics covered in this chapter are:

- *Introduction*
- *Starting Buckling Analysis.*
- *Applying Material.*
- *Defining Fixtures*
- *Applying loads*
- *Defining Connections*
- *Simulating Analysis.*
- *Interpreting results*

INTRODUCTION

Slender models tends to buckle under axial loading. Buckling is defined as the sudden deformation which occurs when the stored membrane (axial) energy is converted into bending energy with no change in the externally applied loads. Mathematically, when buckling occurs, the stiffness becomes singular. The Linearized buckling approach, used here, solves an eigenvalue problem to estimate the critical buckling factors and the associated buckling mode shapes.

A model can buckle in different shapes under different levels of loading. The shape the model takes while buckling is called the buckling mode shape and the loading is called the critical or buckling load. Buckling analysis calculates a number of modes as requested in the Buckling dialog. Designers are usually interested in the lowest mode (mode 1) because it is associated with the lowest critical load. When buckling is the critical design factor, calculating multiple buckling modes helps in locating the weak areas of the model. The mode shapes can help you modify the model or the support system to prevent buckling in a certain mode.

A more vigorous approach to study the behavior of models at and beyond buckling requires the use of nonlinear design analysis codes.

In a laymen's language, if you press down on an empty soft drink can with your hand, not much will seem to happen. If you put the can on the floor and gradually increase the force by stepping down on it with your foot, at some point it will suddenly squash. This sudden scrunching is known as "buckling."

Models with thin parts tend to buckle under axial loading. Buckling can be defined as the sudden deformation, which occurs when the stored membrane (axial) energy is converted into bending energy with no change in the externally applied loads. Mathematically, when buckling occurs, the total stiffness matrix becomes singular.

In the normal use of most products, buckling can be catastrophic if it occurs. The failure is not one because of stress but geometric stability. Once the geometry of the part starts to deform, it can no longer support even a fraction of the force initially applied. The worst part about buckling for engineers is that buckling usually occurs at relatively low stress values for what the material can withstand. So they have to make a separate check to see if a product or part thereof is okay with respect to buckling.

Slender structures and structures with slender parts loaded in the axial direction buckle under relatively small axial loads. Such structures may fail in buckling while their stresses are far below critical levels. For such structures, the buckling load becomes a critical design factor. Stocky structures, on the other hand, require large loads to buckle, therefore buckling analysis is usually not required.

Buckling almost always involves compression. In civil engineering, buckling is to be avoided when designing support columns, load bearing walls and sections of bridges which may flex under load. For example an I-beam may be perfectly "safe" when considering only the maximum stress, but fail disastrously if just one local spot of a flange should buckle! In mechanical engineering, designs involving thin parts in flexible structures like airplanes and automobiles are susceptible to buckling. Even though stress can be very low, buckling of local areas can cause the whole structure to collapse by a rapid series of 'propagating buckling'.

Buckling analysis calculates the smallest (critical) loading required for buckling a model. Buckling loads are associated with buckling modes. Designers are usually interested in the lowest mode because it is associated with the lowest critical load. When buckling is the critical design factor, calculating multiple buckling modes helps in locating the weak areas of the model. This may prevent the occurrence of lower buckling modes by simple modifications.

USE OF BUCKLING ANALYSIS

Slender parts and assemblies with slender components that are loaded in the axial direction buckle under relatively small axial loads. Such structures can fail due to buckling while the stresses are far below critical levels. For such structures, the buckling load becomes a critical design factor. Buckling analysis is usually not required for bulky structures as failure occurs earlier due to high stresses. The procedure to use buckling analysis in SolidWorks Simulation is given next.

Starting the Buckling Analysis

- Open the part on which you want to perform the analysis.
- Click on the **SolidWorks Simulation** button from the **SOLIDWORKS Add-Ins** tab of the **Ribbon** to add **Simulation** tab in the **Ribbon**, if not added already.
- Click on the Down arrow below the **Study Advisor** button and select the **New Study** tool from the drop-down. List of analysis studies that can be performed, will be displayed in the left.
- Select the **Buckling** button and click on the **OK** button from **Study PropertyManager**. The tools related to buckling analysis will be displayed in the **Ribbon**.

Setting the Number of Buckling Modes

- Click on the **Study Properties** tool from the **New Study** drop-down in the **Ribbon**. The **Buckling** dialog box will be displayed as shown in Figure-1.
- Specify the desired number of buckling modes in the **Number of buckling modes** spinner in the dialog box.
- Click on the **OK** button from the dialog box to apply the changes.

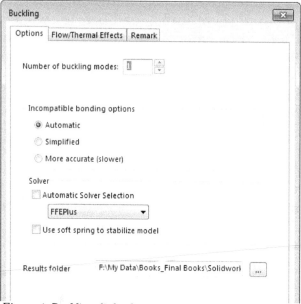

Figure-1. Buckling dialog box

Applying Material

- Click on the **Apply Material** button from the **Ribbon**. The **Material** dialog box will be displayed; refer to Figure-2.
- Select the **1060 Alloy** material under **Aluminium Alloys** node.
- Click on the **Apply** button and then select the **Close** button from the dialog box.

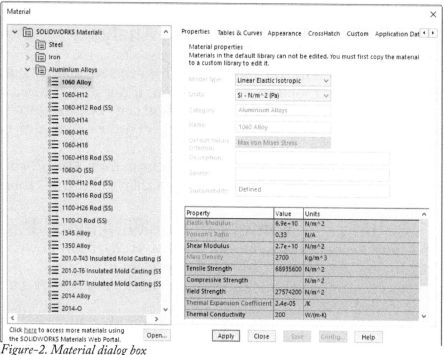

Figure-2. Material dialog box

Applying Fixture

- Click on the down arrow below **Fixtures Advisor** in the **Ribbon**. The list of tools will be displayed.
- Click on the **Fixed Geometry** button from the list. The **Fixture PropertyManager** will be displayed; refer to Figure-3.

Figure-3. Fixture PropertyManager

- Select the face as shown in Figure-4.

Figure-4. Face for fixture

- Click on the **OK** button from the **PropertyManager** to apply the fixture.

Applying the Force

- Click on the down arrow below **External Loads Advisor** in the **Ribbon**. A list of tools will be displayed.
- Click on the **Force** button from the list. The **Force/Torque PropertyManager** will be displayed; refer to Figure-5.

Figure-5. Force Torque PropertyManager

- Select the face as shown in Figure-6.
- Click in the **Force Value** edit box in the **PropertyManager** and specify the value of force as **50000**.

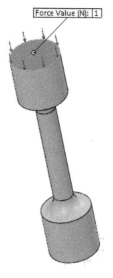

Figure-6. Face for applying force

- Click on the **OK** button from the **PropertyManager** to apply the forces.

Running the Analysis

- Click on the **Run This Study** button from the **Ribbon**. The analysis will start solving.
- After the completion of analysis, the result will be displayed in the modeling area; refer to Figure-7.

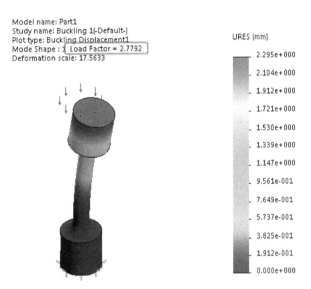

Model name: Part1
Study name: Buckling 1(-Default-)
Plot type: Buckling Displacement1
Mode Shape : 1   Load Factor = 2.7792
Deformation scale: 17.5633

URES (mm)

2.295e+000
2.104e+000
1.912e+000
1.721e+000
1.530e+000
1.339e+000
1.147e+000
9.561e-001
7.649e-001
5.737e-001
3.825e-001
1.912e-001
0.000e+000

Figure-7. Result

- From the above figure, you can check that the load factor for this model is **2.7792**. This load factor represents the factor of safety for the model.
- According to the above value, the model is able to sustain the load for buckling.
- Note that although the part will not fail under the given load but it will be deformed as per the scale so you should also check the non-linear static analysis results.

Note that when buckling analysis is performed only the first mode is important for analysis as this will always be with lowest force at which buckling will occur.

Meaning of different values of load factor are given in Figure-8.

BLF Value	Buckling Status	Remarks
>1	Buckling not predicted	The applied loads are less than the estimated critical loads.
= 1	Buckling predicted	The applied loads are exactly equal to the critical loads. Buckling is expected.
< 1	Buckling predicted	The applied loads exceed the estimated critical loads. Buckling will occur.
-1 < BLF < 0	Bucklin possible	Buckling is predicted if you reverse the load directions.
-1	Buckling possible	Buckling is expected if you reverse the load directions.
< -1	Buckling not predicted	The applied loads are less than the estimated critical loads, even if you reverse their directions.

Figure-8. Buckling load factor

MANUAL SOLUTION OF BUCKLING ANALYSIS

For elastic buckling, we can use the Euler-Column formula of critical buckling load given as

$$P_{cr} = \frac{\pi^2 EI}{L_e^2}$$

Here, P_{cr} is critical buckling load, E is the Elasticity Modulus, I is the moment of inertia, and L_e is the effective length of column which should be considered for analysis.

$I = \pi d^4/64$

There is a relation between effective length and length of column based on fixtures applied on the model. These relations are discussed next.

- If there is fixed condition at one end and free condition on another end of column then effective length will be twice of actual length.
- If both the ends at fixed then effective length will be half of actual length.
- If one end of column is fixed and another is pinned then effective length will be 0.707 x L.
- If both the ends of column are pinned then effective length will be equal to actual length.

Note that while calculating the buckling load, you need to keep units of parameters in same unit system.

If you want to calculate mode shape deflection then for our case, then mathematically it can be given as:

$$y = A(1 - \cos\left(\frac{\pi z}{2L}\right))$$

Elastic Buckling Validation

In elastic buckling, it is assumed that the length of object is very large compared to the size of its base. In elastic buckling theory, we validate this by using Sc and Sm ratios. Sc is the current slender ratio of column and Sm is the minimum slender ratio for stability of object. Generally this type of validation is performed by Civil engineers while constructing columns but the equations are equally good for mechanical design.

Mathematically, these ratios can be defined as:

$$S_c = \frac{L_e}{\sqrt{\frac{I}{A}}} \qquad S_m = \pi\sqrt{\frac{2E}{\sigma_y}}$$

SELF-ASSESSMENT

Q1. In case of buckling, once the geometry of the part starts to deform, it can no longer support even a fraction of the force initially applied. (T/F)

Q2. Buckling occurs at relatively low stress values for what the material can withstand. (T/F)

Q3. Buckling analysis is usually not required for bulky structures as failure occurs earlier due to high stresses. (T/F)

Q4. The load factor in Buckling analysis results represents the factor of safety for the model. (T/F)

Q5. If the load factor is 1 then our model is safe from buckling. (T/F)

FOR STUDENT NOTES

Answer to Self-Assessment:
1. T, 2. T, 3. T, 4. F, 5. F

Chapter 11

Fatigue Analysis

Topics Covered

The major topics covered in this chapter are:

- *Introduction*
- *Starting Fatigue Analysis.*
- *Selecting the previous analysis*
- *Defining the number of cycles*
- *Simulating Analysis.*
- *Interpreting results*

INTRODUCTION

Repeated loading and unloading weakens objects over time even when the induced stresses are considerably less than the allowable stress limits. This phenomenon is known as fatigue. Each cycle of stress fluctuation weakens the object to some extent. After a number of cycles, the object becomes so weak that it fails. Fatigue is the prime cause of the failure of many objects, especially those made of metals. Examples of failure due to fatigue include, rotating machinery, bolts, airplane wings, consumer products, offshore platforms, ships, vehicle axles, bridges, and bones.

Linear and nonlinear structural studies do not predict failure due to fatigue. They calculate the response of a design subjected to a specified environment of restraints and loads. If the analysis assumptions are observed and the calculated stresses are within the allowable limits, they conclude that the design is safe in this environment regardless of how many times the load is applied.

Results of static, nonlinear, or time history linear dynamic studies can be used as the basis for defining a fatigue study. The number of cycles required for fatigue failure to occur at a location depends on the material and the stress fluctuations. This information, for a certain material, is provided by a curve called the S-N curve.

Stages of Failure Due to Fatigue

Failure due to fatigue occurs in three stages:

Stage 1 One or more cracks develop in the material. Cracks can develop anywhere in the material but usually occur on the boundary faces due to higher stress fluctuations. Cracks can occur due to many reasons. Imperfections in the microscopic structure of the materials and surface scratches caused by tooling or handling are some of them.

Stage 2 Some or all the cracks grow as a result of continued loading.

Stage 3 The ability of the design to withstand the applied loads continue to deteriorate until failure occurs.

Fatigue cracks start on the surface of a material. Strengthening the surfaces of the model increases the life of the model under fatigue events.

From the above theory, you can find out that the main player in Fatigue analysis is the properties of materials rather the S-N curves.

The procedure to perform the Fatigue analysis is given next.

Starting Fatigue Analysis

- Open a part on which you have already performed Static or dynamic analysis. (We have opened the file used in Chapter 4 of this book; refer to Figure-1.)

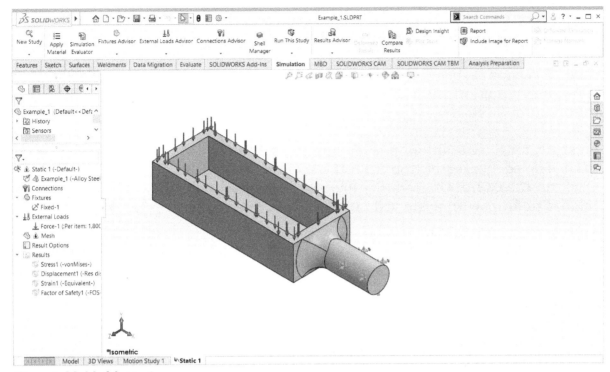

Figure-1. Model of chapter 3

- Click on the down arrow below **New Study** button, the list of tools will be displayed.
- Click on the **New Study** button from the list. The **Study PropertyManager** will be displayed.
- Click on the **Fatigue** button from the list and then click on the **Variable amplitude history data** button from the **Options** rollout of the **PropertyManager**; refer to Figure-2.

Note that in this example, we will study the Variable amplitude history data fatigue study but in SolidWorks simulation, we can perform four type of fatigue analyses viz. **Constant amplitude events with defined cycles**, **Variable amplitude history data**, **Harmonic- fatigue of sinusoidal loading**, and **Random vibration-fatigue of random vibration**. The procedure of all the studies is almost the same.

In case of **Constant amplitude events with defined cycles**, the load being applied on the object is constant and we specify the number of loading cycles to test the fatigue.

In case of **Variable amplitude history data**, the load applied on object varies with time. This variation can be defined in the fatigue study itself.

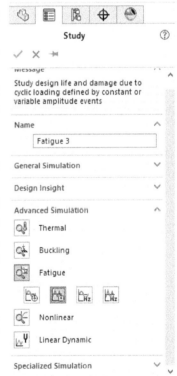

Figure-2. Variable amplitude history data button

In case of **Harmonic- fatigue of sinusoidal loading**, the dynamic load is applied in the form of sinusoidal curve. For this type of fatigue study, we require the data from studies performed under Linear Dynamic category.

In case of **Random vibration-fatigue of random vibration**, we require the data from random vibration study. To perform this study, you need to perform the random vibration study first.

- To study for a constant amplitude fatigue, select the **Constant amplitude events with defined cycles** button from the **Options** rollout.
- Click on the **OK** button from the **PropertyManager** to start the analysis. The tools related to fatigue analysis will be displayed; refer to Figure-3.

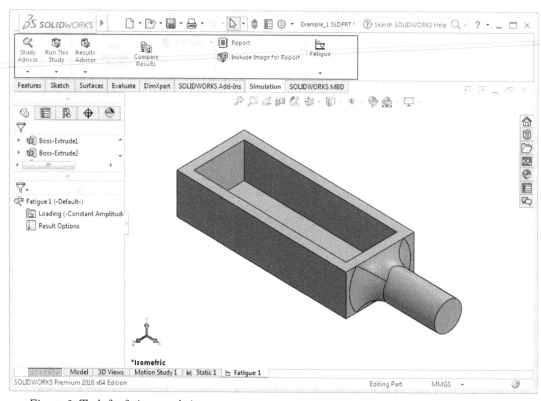

Figure-3. Tools for fatigue analysis

Initiating Fatigue Analysis

- Click on the down arrow below **Fatigue** in the **Ribbon**. The **Add Event** tool will be displayed.
- Click on the **Add Event** tool. The **Add Event (Variable) PropertyManager** will be displayed; refer to Figure-4.

Figure-4. Add Event PropertyManager

- By default, Static 1 study is added in the event list.
- Click in the **Number of repeats** spinner in the **Options** rollout and specify the number of repetitions as **300** for the load curve being generated in the next segment.

Now, we need to specify the time curve for the fatigue analysis.

- Click on the **Get Curve** button from the **PropertyManager**. The **Load History Curve** dialog box will be displayed; refer to Figure-5.

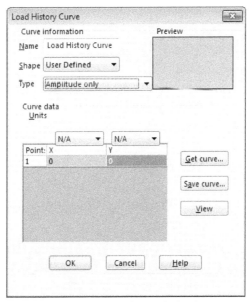

Figure-5. Load History Curve dialog box

- Click in the **Name** edit box and specify the desired name of the curve.

- Click on the **Type** drop-down and select the **Time & amplitude** option from the list displayed.
- The table of time versus amplitude is displayed in the **Curve data** area.
- Double-click on the **1** button under the **Point** column. The next point will be added in the table; refer to Figure-6.

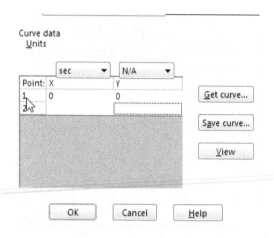

Figure-6. Curve data after adding one more point

- Similarly, you can add more points.
- Enter the data related to amplitude in the table; refer to Figure-7.

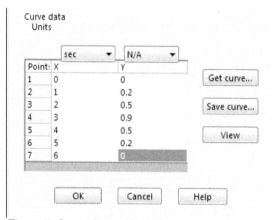

Figure-7. Curve data table with custom values

- Click on the **OK** button from the dialog box.

- Click on the **OK** button from the **PropertyManager** to add the event.

Applying or Editing S-N data

- Right-click on the name of the model in the **Analysis Manager** and select the **Apply/Edit Fatigue data** button; refer to Figure-8.

Figure-8. Apply Edit Fatigue Data button

- The **Fatigue SN Curves** tab in **Materials** dialog box will be displayed; refer to Figure-9.

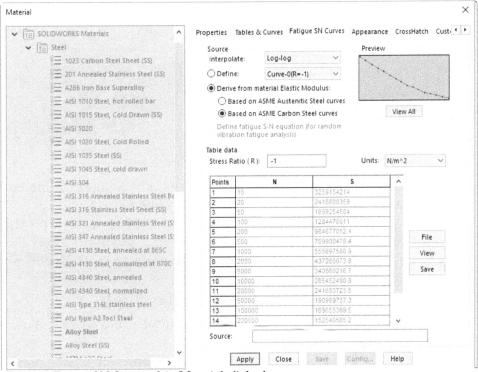

Figure-9. Fatigue SN Curve tab in Materials dialog box

- If you have the table of parameters then specify the data as done in case of **Curve Data**. In this case, we will use the fatigue data of **ASME Austenitic Steel**.
- Click on the **Derive from material Elastic Modulus** radio button from the **Source** area in the dialog box. The data will be imported automatically in the table.
- Click on the **Apply** button and then **Close** button from the dialog box.

Modifying the Static Analysis earlier performed

- Click on the **Static 1** tab in the bottom bar of **SolidWorks**; refer to Figure-10. The static analysis will open; refer to Figure-11.

Figure-10. Static 1 tab

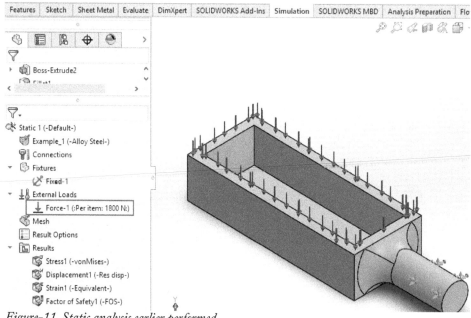

Figure-11. Static analysis earlier performed

- Double click on the **Force-1** node in the **Analysis Manager**; refer to Figure-11. The **Force/Torque PropertyManager** will be displayed.
- Change the force value in the **PropertyManager** to **1800** and click on the **OK** button from the **PropertyManager**.
- Click on the **Run This Study** button from the **Ribbon** to update the analysis.

Running the Fatigue Analysis

- Click on the **Fatigue 1** tab from the bottom bar of SolidWorks. You will switch back to fatigue analysis.
- Click on the **Run This Study** button from the **Ribbon**.
- Double-click on the **Results2 (-Life-)** node in the **Analysis Manager**. The results will be displayed as shown in Figure-12.

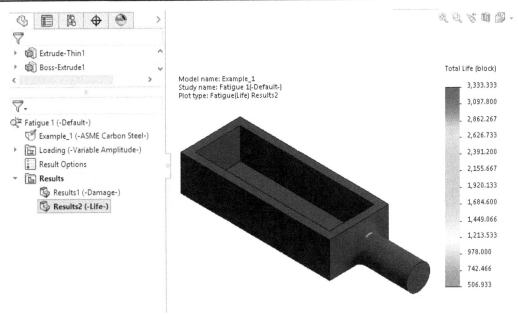

Figure-12. Results

- From the results, we can find out that the part will fail on **506th** cycle at the location painted in red.
- The red location is below the green marked area, so you need to rotate the model; refer to Figure-13.

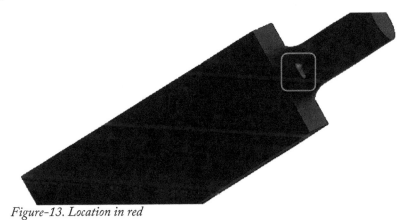

Figure-13. Location in red

Changing Properties of Fatigue Analysis

- Right-click on the **Fatigue 1 (-Default-)** node in the **Analysis Manager**. The options related to fatigue analysis will be displayed.
- Click on the **Properties** button from the list displayed. The **Fatigue** dialog box will be displayed; refer to Figure-14.

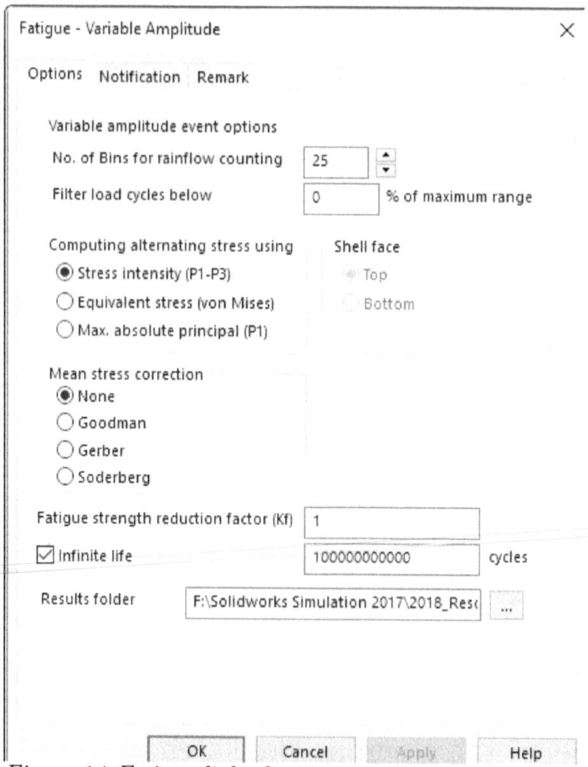

Figure-14. Fatigue dialog box

- The **No. of Bins for rainflow counting** spinner is used to specify the number of divisions of load frequency for specified amplitude.
- The options in the **Computing alternating stress using** area are used to specify the analysis result to be used for testing fatigue.
- The options in the **Mean stress correction** area are used to set the correction on the basis of material properties. **Described as**:
  None: No correction.
  Goodman method: Generally suitable for brittle materials.
  Gerber method: Generally suitable for ductile materials.
  Soderberg method: Generally the most conservative.
- When meshing is done, SolidWorks creates two type of shell elements: Top face shell elements and Bottom face shell elements. This elements can be distinguished by their color. The Shell face area of the dialog box allows you to select the faces that you want to use for performing fatigue analysis.
- The **Fatigue strength reduction factor (Kf)** edit box is used to specify the value of fatigue reduction factor for allowing environment factor in analysis. Use this factor, between 0 and 1, to account for differences in test environment used to generate the S-N curve and the actual loading environment. The program divides the alternating stress by this factor before reading the corresponding number of cycles from the S-N curve. This is equivalent to reducing the number of cycles that cause failure at a certain alternating stress. Fatigue handbooks suggest numeric values for the fatigue strength reduction factor.
- The **Filter load cycles below** edit box forces the system to filter out load cycles with ranges smaller than the specified percentage of the maximum range. For example, if you specify 5%, the program ignores cycles with load ranges less than 5% of the maximum range of the load history. Use this parameter to filter out noise from measuring devices.

RUNNING CONSTANT AMPLITUDE EVENTS
ON THE MODEL

- Start a new study of Fatigue by using the **Constant amplitude** button from the **Study PropertyManager**; refer to Figure-15. Click on the OK button to start the analysis.

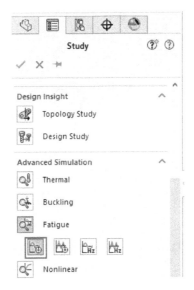

Figure-15. Constant amplitude button

- Click on the **Add Event** button from the **Fatigue** drop-down in the **Ribbon**. The **Add Event(Constant) PropertyManager** will be displayed as shown in Figure-16.

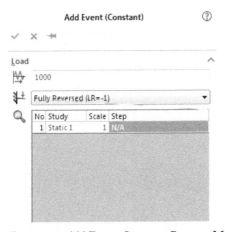

Figure-16. Add Event Constant PropertyManager

- Make sure the number of cycles are **1000** in the **Cycles** edit box. Click on the **OK** button from **PropertyManager**.
- Click on the **Run This Study** button from the **Ribbon**. The results of constant loading will be displayed as shown in Figure-17.

Model name:Example_1
Study name:Fatigue 2(-Default-)
Plot type: Fatigue(Life) Results2

Total Life (cycle)

- 1.000e+006
- 9.170e+005
- 8.340e+005
- 7.510e+005
- 6.680e+005
- 5.850e+005
- 5.020e+005
- 4.190e+005
- 3.360e+005
- 2.530e+005
- 1.700e+005
- 8.701e+004
- 4.007e+003

Figure-17. Result of constant loading fatigue

SELF-ASSESSMENT

Q1. Linear and nonlinear structural studies do not predict failure due to fatigue. (T/F)

Q2. Results of static, nonlinear, or time history linear dynamic studies can be used as the basis for defining a fatigue study. (T/F)

Q3. The number of cycles required for fatigue failure to occur at a location depends on the material not on the stress fluctuations. (T/F)

Q4. Cracks due to fatigue can develop anywhere in the material but usually occur on the boundary faces due to higher stress fluctuations. (T/F)

Q5. Strengthening the surfaces of the model does not affect the life of the model under fatigue events. (T/F)

Answer to Self-Assessment:
1. T, 2. T, 3. T, 4. T, 5. F

Chapter 12

Drop-Test

Topics Covered

The major topics covered in this chapter are:

- *Introduction*
- *Starting Drop-Test*
- *Applying Material*
- *Setting drop parameters*
- *Simulating Analysis.*
- *Interpreting results*

INTRODUCTION

In a drop test analysis, the time varying stresses and deformations due to an initial impact of the product with a rigid or flexible planar surface (the floor) are calculated. As the product deforms, secondary internal and external impacts are also calculated, locating critical weaknesses or failure points, as well as stresses and displacements. Drop test analysis using SolidWorks Simulation enables you to visualize the elastic stress wave propagating through the system so that the correct assembly methods are used.

The maximum "g force" experienced by individual components is one of the primary unknowns before a drop test. This is a critical parameter since many electronic and mechanical components are not rated for further use above a specified maximum g force. Using drop test analysis with SolidWorks Simulation, designers and engineers can measure the time varying accelerations (g force) at any location within the product, providing critical design information and reducing the number of physical tests needed. A design team can easily verify performance while they design and select the correct materials, component shape, and fixture methods to ensure that critical components stay within their "max g force" limits.

STARTING DROP TEST

- Open the part on which you want to perform the drop-test.
- Click on the **SolidWorks Simulation** button from the **SOLIDWORKS Add-Ins** tab of the **Ribbon** to add **Simulation** tab in the **Ribbon**, if not added already.
- Click on the Down arrow below the **Study Advisor** button and select the **New Study** tool from the drop-down. List of analysis studies will be displayed in the **Study PropertyManager**.
- Click on the **Drop Test** button from the list and click on the **OK** button from the **PropertyManager**. The tools related to drop-test will be displayed; refer to Figure-1.

Applying Material

- Right-click on name of the model i.e. Scale Base. A shortcut menu will be displayed.
- Click on **Apply/Edit Material** button from the shortcut menu. The **Material** dialog box will be displayed.
- Select the **ABS PC Material** from the **Plastic** category in the left; refer to Figure-2.

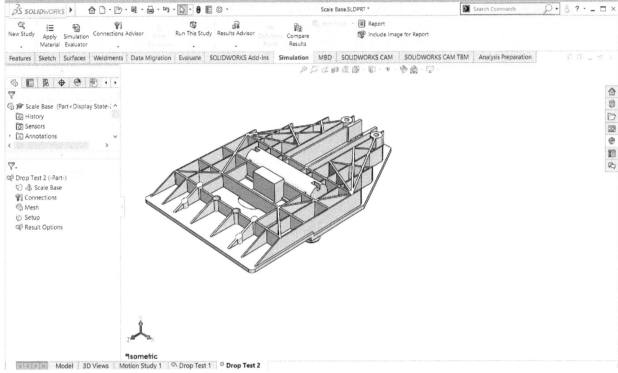

Figure-1. Tools related to drop test

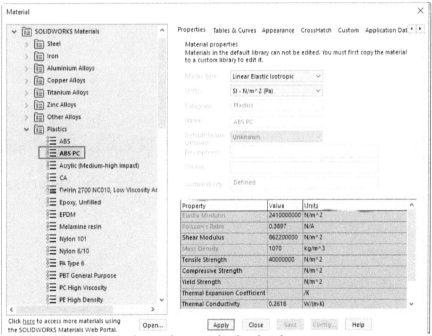

Figure-2. Material dialog box with material to be selected

- Click on the **Apply** button and then click on the **OK** button from the dialog box.

If you are working on an assembly then you can apply the contacts as discussed in earlier chapters.

Creating Mesh

- Right-click on the **Mesh** in the **Analysis Manager**. A shortcut menu will be displayed.
- Click on the **Simplify Model for Meshing** button from the shortcut menu. The **Simplify** task pane will be displayed; refer to Figure-3.

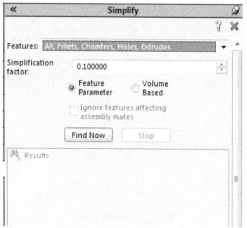

Figure-3. Simplify taskpane

- Click on the **Find Now** button to find the areas that create complexity in meshing. The complex areas are displayed below **Results** in the **Simplify** task pane; refer to Figure-4.

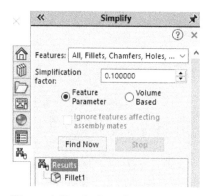

Figure-4. Results in task pane

- Close the **Simplify** task pane and select all the elements by holding **CTRL** key while selecting from the **FeatureManager**.
- Right-click on any of the selected element and select the **Suppress** button; refer to Figure-5.

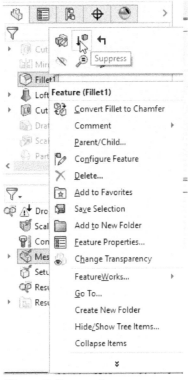

Figure-5. Suppress button

- On doing so, the selected elements will be suppressed.
- Again, right-click on **Mesh** in the **Analysis Manager** and select the **Create Mesh** button. The **Mesh PropertyManager** will be displayed; refer to Figure-6.

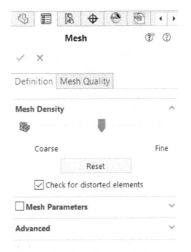

Figure-6. Mesh PropertyManager

- Set the mesh density as per your system capabilities and accuracy requirement.
- Click on the **OK** button from the **PropertyManager** to create the mesh.

Parameters for Drop

- Right-click on the **Setup** node from the **Analysis Manager**. A shortcut menu will be displayed; refer to Figure-7.
- Click on the **Define/Edit** button from the menu. The **Drop Test Setup PropertyManager** will be displayed; refer to Figure-8.

Figure-7. Setup shortcut menu

Figure-8. Drop Test Setup
PropertyManager

- If you know the velocity of the object being dropped then click on the **Velocity at impact** radio button from the **Specify** rollout in the **PropertyManager**. The **Velocity at Impact** rollout will be displayed in the **PropertyManager**; refer to Figure-9.

Figure-9. Velocity at Impact rollout

- Select the reference for velocity direction and select the **Reverse** button adjacent to the selection box in the rollout.
- Enter the desired velocity in the edit box in **Velocity at Impact** rollout.

- If you know height of the object from where it is being dropped, then click on the **Drop height** radio button. The **Height** rollout will be displayed; refer to Figure-10. Note that the **Drop height** radio button is selected by default.

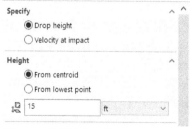

Figure-10. Height rollout

- Select the **From centroid** radio button if you want to specify the drop height from centroid of the model or select the **From lowest point** radio button if you want to specify the height from the lowest point of the model.
- Specify the drop height in the edit box in **Height** rollout.

- Click in the Pink selection box in the **Gravity** rollout and select the plane/face/ edge for specifying the gravity direction reference.
- Select the face for direction reference; refer to Figure-11.

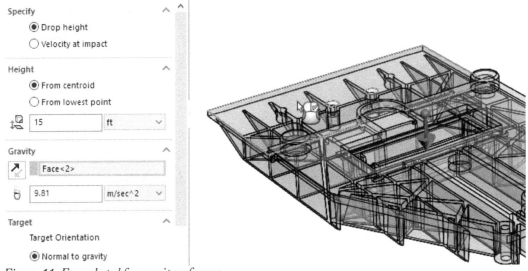

Figure-11. Face selected for gravity reference

- You can change the gravitational acceleration as per your requirement by using the **Gravity Magnitude** edit box in the **Gravity** rollout, if required.
- Select the **Normal to gravity** radio button in the **Target** rollout if the target plane (ground) is normal to gravity direction.
- Select the **Parallel to ref. plane** radio button and select the reference plane for specifying the ground orientation as per your requirement.
- You can specify the friction coefficient for selected material by using the **Friction Coefficient** edit box, if required.
- If the floor is tough then select the **Rigid target** radio button. If your floor is soft then select the **Flexible target** radio button and specify the related parameters in the **Stiffness and Thickness** rollout displayed in the **PropertyManager**.
- The edit box in **Contact Damping** rollout is used to specify the damping ratio for collision.
- After specifying the desired parameters, click on the **OK** button from the **PropertyManager** to apply the settings.

Running the Analysis

- Click on the **Run This Study** button from the **Ribbon**. System will start solving the analysis.
- After the analysis is solved, the results will be displayed; refer to Figure-12.

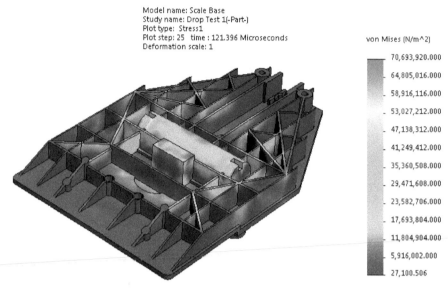

Figure-12. Results

- You can create more plots by right-clicking on **Results** node in the **Analysis Manager** and selecting the desired result option.

Chapter 13

Pressure Vessel Design and Design Study

Topics Covered

The major topics covered in this chapter are:

- *Starting Pressure Vessel Design.*
- *Inclusion of Analyses*
- *Simulating Design Study*
- *Starting Design Study*
- *Specifying Variables*
- *Setting Constraints*
- *Setting Goals*
- *Evaluating results*

INTRODUCTION

Pressure vessels are containers that store the pressurized fluids. A pressure vessel can be of any practical shape but generally spherical, cylindrical, and conical shapes are employed. The most important question for pressure vessel designer is whether the designed vessel will sustain the required conditions or not. SolidWorks Simulation provides a separate option to perform this study named **Pressure Vessel Design** study.

In a **Pressure Vessel Design** study, you can combine the results of static studies with the desired factors. Each static study has a different set of loads that produce corresponding results. These loads can be dead loads, live loads (approximated by static loads), thermal loads, seismic loads, and so on. The **Pressure Vessel Design** study combines the results of the static studies algebraically using a linear combination or the square root of the sum of the squares (SRSS).

Before we start performing the **Pressure Vessel Design** study, there must be two or more static analyses performed on the vessel earlier. In the next example, we have performed two static analyses and one thermal analysis on the model. Details of these analyses are given as follows:

Thermal Analysis 1

Parameters for thermal analysis are given in Figure-1.

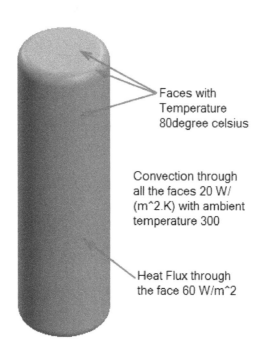

Faces with
Temperature
80degree celsius

Convection through
all the faces 20 W/
(m^2.K) with ambient
temperature 300

Heat Flux through
the face 60 W/m^2

Figure-1. Parameters for thermal analysis

You can perform the analysis as per the above figure for your practice. Note that the analysis is already performed in the part file supplied in the Resource kit.

Static Analysis 1 & 2

In the Figure-2, the parameters for static analysis 1 are given and in the Figure-3 parameters for static analysis 2 are given.

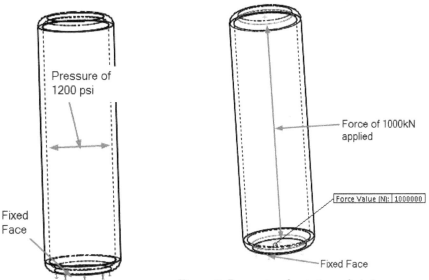

Figure-3. Parameters for static analysis 2

Figure-2. Parameters for static analysis 1

STARTING PRESSURE VESSEL DESIGN STUDY

- Open the file on which you have performed the above given analyses.
- Click on the **Simulation** tab in the **Ribbon** and click on the down arrow below **Study Advisor**. List of tools will be displayed.
- Click on the **New Study** button and select the **Pressure Vessel Design** button from the **PropertyManager**.
- Click on the **OK** button from the **PropertyManager** to display the tools related to **Pressure Vessel Design Study**; refer to Figure-4.

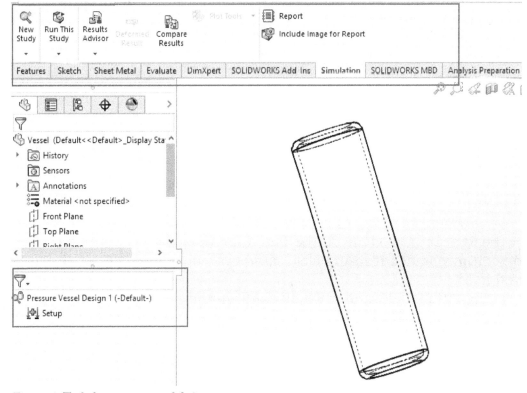

Figure-4. Tools for pressure vessel design

Setup for Design Study

- Right-click on the **Setup** node in the **Analysis Manager**. A shortcut menu will be displayed; refer to Figure-5.
- Click on the **Define/edit** button. The **Result Combination Setup PropertyManager** will be displayed; refer to Figure-6.

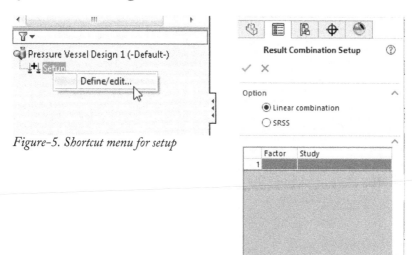

Figure-5. Shortcut menu for setup

Figure-6. Result Combination Setup PropertyManager

There are two ways to combine various analyses results to study pressure vessel design; Linear combination and SRSS. Select the **Linear combination** radio button to linearly add the results of various studies to display final result. If you select the **SRSS** radio button then the square root of the sum of the squares study results will be used to represent the final result.

- Click in the field below **Study** column, the field will be converted to drop-down.
- Click on the down arrow of drop-down. The list of static analyses studies will displayed; refer to Figure-7.

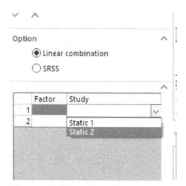

Figure-7. List of studies

- Select the **Static 1** study from the list. **1** factor will automatically appear in the **Factor** column adjacent to the study and one more row will be added below the selected field.
- Click on the empty field under **Study** column and select the **Static 2** study from the drop-down as discussed earlier.
- Click on the **OK** button from the **PropertyManager**.

Running the Design Study

- Click on the **Run This Study** button in the **Ribbon**. The combining process of two studies will start.
- After the study is complete, the results will be displayed in the **Result** node of the **Analysis Manager**; refer to Figure-8.

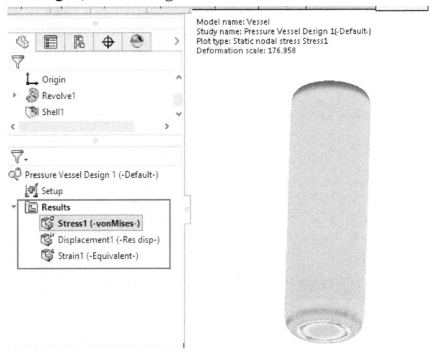

Figure-8. Results after combination

- Right-click on the **Results** node and select the **Define Factor of Safety Plot** button. The **Factor of Safety PropertyManager** will be displayed; refer to Figure-9.

Figure-9. Factor of Safety PropertyManager

- Click on the **OK** button from the **PropertyManager**. The factor of safety plot will be created in the modeling area; refer to Figure-10.

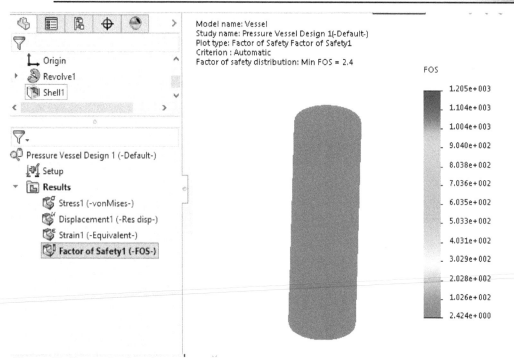

Figure-10. Factor of safety plot

DESIGN STUDY

Design Study is used to perform the optimization of the model. There are various parameters of an object that decide whether the model under testing will sustain or fail under the given condition. Out of all the parameters of a model, there are a few parameters of our concern that we want to use for satisfying our objective. In a Design Study, our objective can be:

- Reducing the mass of the model.
- Decreasing the total manufacturing cost of the model.
- Increasing the strength of the model and so on.

To achieve these objective, we need to manipulate with dimensions, materials, process and so on. In SolidWorks Simulation, Design Study is used to perform an optimization or evaluate specific scenarios of your design.

You can plot the updated bodies and the calculated results for different iterations or scenarios by selecting their columns on the Results View tab.

You tackle a vast number of problems using a design study.

You can:

- Define multiple variables using any simulation parameter, or driving global variable.
- Define multiple constraints using sensors.
- Define multiple goals using sensors.
- Analyze models without simulation results. For example, you can minimize the mass of an assembly with the variables, density and model dimensions, and the constraint, volume.

- Evaluate design choices by defining a parameter that sets bodies to use different materials as a variable.

The procedure to use the **Design Study** is given next.

Before starting the design study, it is important to understand that the design study is a kind of macro which performs a specific type of analysis with a varying values of a specific parameters. For example, we have performed a static analysis on a tube under a specific pressure and we find that its factor of safety is 64. Then, we will perform a design study to find out the thickness of tube which is optimum i.e. factor of safety is greater than 1 and thickness of tube is minimum possible. In the next example, we have performed such a test on the model.

Starting Design Study

- Open the model on which you have performed the relevant analysis. In this example, we have performed a static analysis and we have found the results as given in Figure-11.

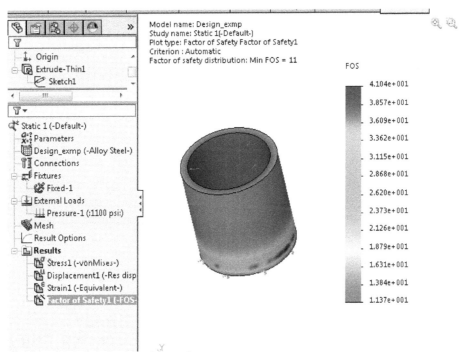

Figure-11. Factor of safety of example

- From the above figure, you can find that the factor of safety of the model under the given test is **11**.
- If we check the dimensions then the thickness of the model is **7** mm; refer to Figure-12.
- To check the dimension of the model, you need to click on the **Measure** button from **Evaluate** tab and select the desired parameter. The dimension will be displayed.

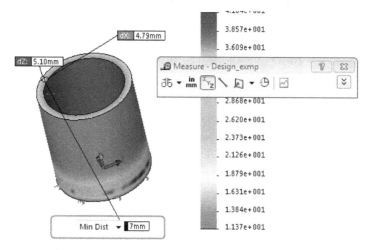

Figure-12. Thickness of tube

- Now, click on the **Simulation** tab in the **Ribbon** to return to simulation study. The tools related to analysis will display.
- Click on the down arrow below **Study Advisor** and select the **New Study** tool from the list of tools displayed. The **Study PropertyManager** will be displayed.
- Click on the **Design Study** button and click on the **OK** button from the **PropertyManager**.
- The **Design Study** pane will be displayed; refer to Figure-13.

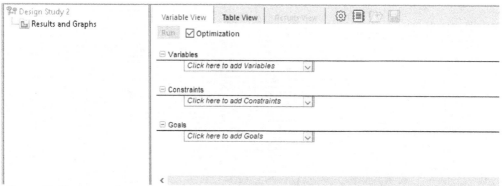

Figure-13. Design study pane

Setting options for Design Study

- Click on the **Design Study Options** button from the **Design Study** pane; refer to Figure-14. The **Design Study Properties PropertyManager** will be displayed; refer to Figure-15.

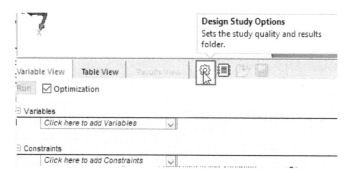

Figure-14. Design Study Options button

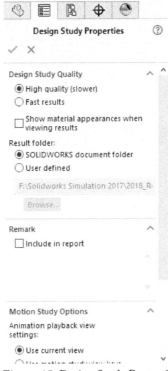

Figure-15. Design Study Properties PropertyManager

- Select the required radio button and set the desired parameters like; for fast results, click on the **Fast results** radio button. After specifying the parameters, click on the OK button from **Design Study Properties PropertyManager**. You will be returned to the **Design Study** tab.

Specifying parameters that are to be changed

- Click on the down arrow under the **Variables** node. A list of variable options will be displayed; refer to Figure-16.

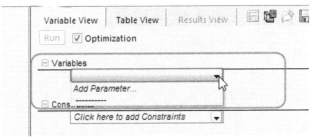

Figure-16. Variables drop-down

- Click on the **Add Parameter** option from the drop-down list. The **Parameters** dialog box will be displayed and you will be prompted to select a dimension that is to be varied; refer to Figure-17.

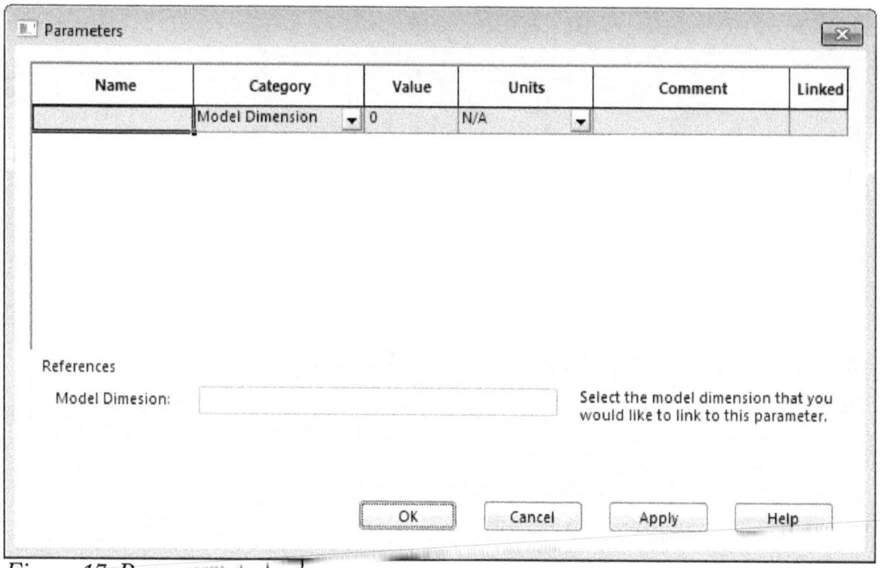

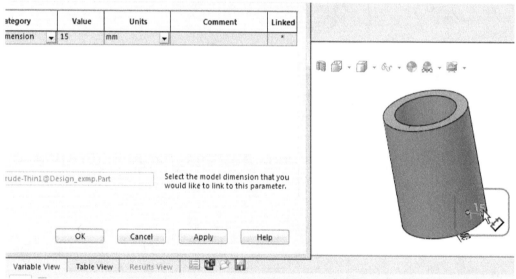

Figure 17. Parameters dialog box

- Click on the dimension of the model that you want to make variable; refer to Figure-18. The selected dimension will be added under the **Value** column in the **Parameters** dialog box.

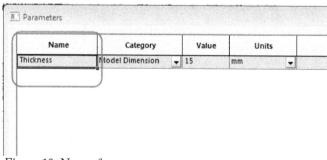

Figure-18. Dimension selected as variable

- Click in the field below **Name** column in the dialog box and specify the name as **Thickness**; refer to Figure-19.

Figure-19. Name of parameter

- If you want to specify other parameters, then click in the name field of second row and follow the same procedure. You can also specify other parameters in spite of dimensions. To do so, click on the down arrow next to **Model Dimension** in **Category** column and select the desired category. There are four categories from which you can specify the variables: **Model Dimension**, **Global Variable**, **Simulation** data, and **Material** data.
- After specifying the variables, click on the **OK** button from the dialog box. The variable will be added in the **Design Study** pane and its range will be displayed; refer to Figure-20.

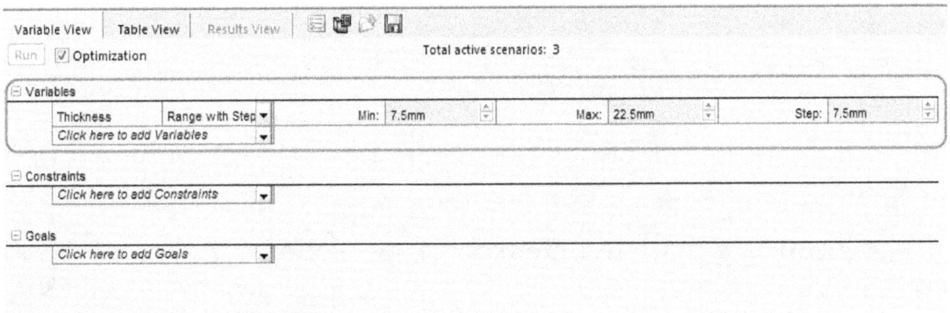

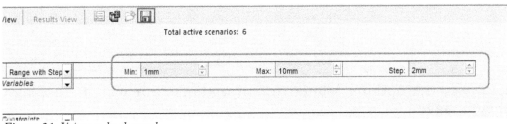

Figure-20. Design variable with range

- We know that the thickness of our model is **7 mm** with **Factor of Safety** as **11**. Now, we need to move below this thickness value to get the optimum thickness of the tube.
- Change the value of **Min**, **Max**,and **Step** as shown in Figure-21 by using the spinners.

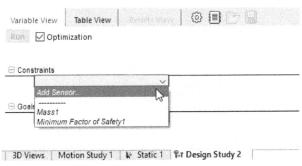

Figure-21. Values to be changed

Specifying restrictions for changes

- Click on the down arrow under the **Constraints** node in the **Design Study** pane. A list of options will be displayed; refer to Figure-22.

Figure-22. Add Sensor button

- Click on the **Add Sensor** option from the list. The **Sensor PropertyManager** will be displayed; refer to Figure-23.

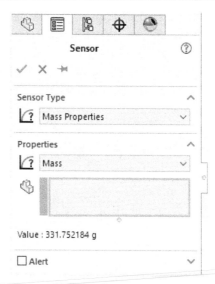

Figure-23. Sensor PropertyManager

- Click on the drop-down in the **Sensor Type** rollout. A list of options will be displayed.
- Select the **Simulation Data** option from the list; refer to Figure-24.

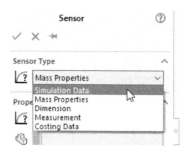

Figure-24. Simulation Data option

- On doing so, the options related to simulation data will be displayed in the **PropertyManager**.
- Click on the **Results** drop-down in the **Data Quantity** rollout and select the **Factor of Safety** option; refer to Figure-25.

Figure-25. Factor of safety option

- Click on the **OK** button from the **Sensor PropertyManager** to add the **Factor of Safety** constraint. The constraint will be added in the **Design Study** pane; refer to Figure-26.

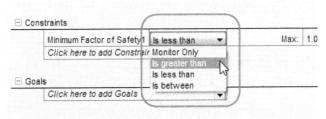

Figure-26. Factor of safety in Design Study pane

- Click on the down arrow next to **is less than** field and select the **Is greater than** option; refer to Figure-27.

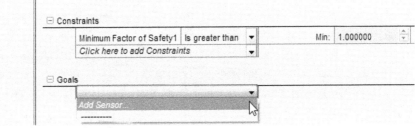

Figure-27. Option to be selected

Setting Goal for the study

- Click on the down arrow under the **Goals** node from **Design Study** . A list of options will be displayed; refer to Figure-28.

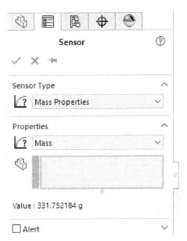

Figure-28. Goals drop down

- Click on the **Add Sensor** option. The **Sensor PropertyManager** will be displayed as discussed earlier; refer to Figure-29.

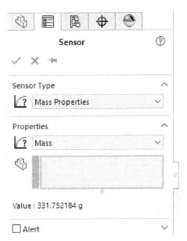

Figure-29. Sensor PropertyManager for goals

- Click on the **OK** button from the **PropertyManager** to set the mass to be minimized. The **Mass** will be added as goal in the **Design Study** pane; refer to Figure-30.

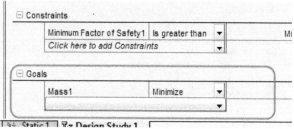

Figure-30. Goal added in Design Study

Running the Study

We have specified all the necessary parameters for the study and ready for the analysis.

- Click on the **Run** button in the **Design Study** pane, the design study will start.
- After the study is solved, the results will be displayed as shown in Figure-31.

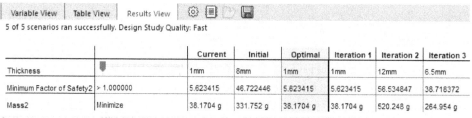

		Current	Initial	Optimal	Iteration 1	Iteration 2	Iteration 3
Thickness		1mm	8mm	1mm	1mm	12mm	6.5mm
Minimum Factor of Safety2	> 1.000000	5.623415	46.722446	5.623415	5.623415	56.534847	38.718372
Mass2	Minimize	38.1704 g	331.752 g	38.1704 g	38.1704 g	520.248 g	264.954 g

Figure-31. Results of design study

- The green highlighted box displays the optimum result for the analysis and its related parameters used for optimization.
- Click on the **Save** button to save the study for future references.
- Click on the **Report** button to generate a word report of the current study; refer to Figure-32. The **Report Options** dialog box will be displayed; refer to Figure-33.

Figure-32. Report button

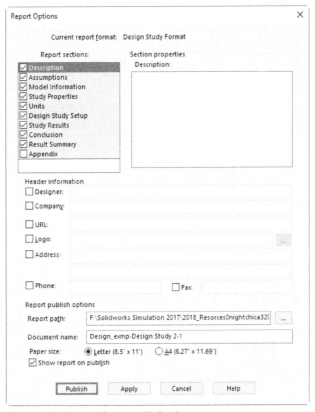

Figure-33. Report Options dialog box

- Enter the desired details of current study in **Report Options** dialog box and click on the **Publish** button. The report will be generated and displayed in Microsoft Word file.

You can specify more variables for changing the design and test the results for practice.

FOR STUDENT NOTES

FOR STUDENT NOTES

Chapter 14

Topology Study

Topics Covered

The major topics covered in this chapter are:

- *Introduction to Topology Study*
- *Applying Material*
- *Applying Fixtures*
- *Applying Force/Torque*
- *Goals and Constraints*
- *Manufacturing Controls*

TOPOLOGY STUDY

Generally, Topology is the mathematical study of the properties that are preserved through deformations, stretching of object, and twisting. In topology, tearing is not allowed. In this type of study, a circle is equivalent to ellipse by stretching and similarly sphere is equivalent to ellipsoid.

Added in Solidworks 2018, Topology study perform structural topology analysis to discover new minimal design alternatives by reducing the costly prototypes and material costs. The Topology study in SolidWorks Simulation simplifies structural investigation of models with a goal driven approach to mathematically alter the stiffness of the meshed geometry. Assuming linear static loading on the model, designers and engineers can reduce mass of the component and can attain the required factor of safety. The results of topology study show which regions of a component can be removed without affecting the strength and stiffness of component. In Topology analysis, the engineers and designers are able to:

- Set a target mass reduction
- Limit material removal by limiting maximum component displacement.
- Enforce manufacturing constraints.

The results of topology study can be used as a input for a 3D printer or as a guide for material removal of a part.

Topology study is generally used to explore design iterations of a component that satisfy a given optimization goal and geometric constraints. When designer is designing new model or part, if the part is too large and occupy much space then it can limit the creativity of designer. Here, topology study allows you to search for an alternative or a smaller shape model to meet designer requirement. It can help you to reduce the size or material of part without reducing the stiffness of model.

STARTING TOPOLOGY STUDY

- Open the part on which you want to perform the Topology Study.
- Click on the **SolidWorks Simulation** button from the **SOLIDWORKS Add-Ins** tab of the **Ribbon** to add **Simulation** tab in the **Ribbon**, if not added already.
- Click on the Down arrow below the **Study Advisor** button and select the **New Study** tool from the drop-down. List of analysis studies will be displayed in the **Study PropertyManager**.
- Click on the **Topology Study** button from the list and click on the **OK** button from the **PropertyManager**. The **Topology Study PropertyManager** will be displayed; refer to Figure-1.

Figure-1. Topology Study Property Manager

- If you are working on a part file then the options related to **Topology Study** will be displayed in the **Task List - Topology Study** pane instead of **Topology Study PropertyManager**; refer to Figure-2. In our case, we are working on an assembly file. After selecting the **Topology Study** button from the list, the **Topology Study PropertyManager** will be displayed to select a part of assembly to work with.

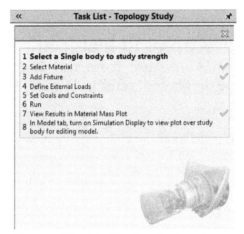

Figure-2. Task List Topology Study task pane

- Click on the component from assembly on which you want to do a topology study. The selected component will be displayed in **Maximum Sized Model** selection box. After selecting the desired model, click on the **OK** button from **Topology Study PropertyManager**; refer to Figure-3. The options related to **Topology Study** will be displayed along with the model in **Task List - Topology Study** task pane will be displayed.

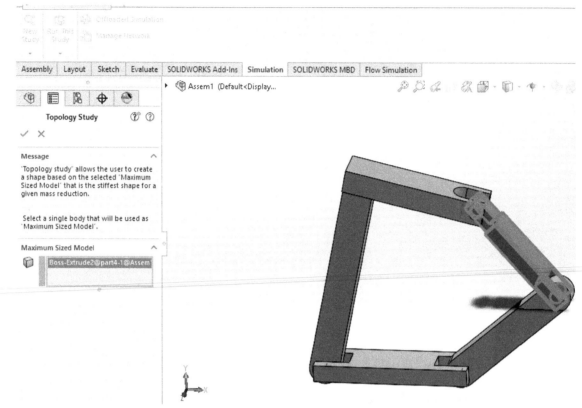

Figure-3. Selection of component from assembly

Applying Material

- Click on the **Apply Material** button from the **Simulation** tab of **Ribbon**. The **Material** dialog box will be displayed; refer to Figure-4. Select the desired material from the material list.

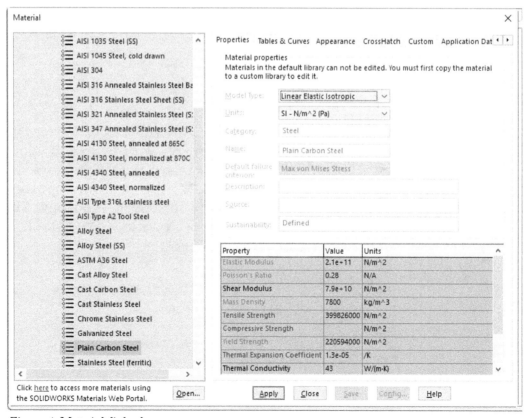

Figure-4. Material dialog box

- Click on the **Apply** button from **Material** dialog box and click on the **Close** button. The selected material will applied on the component.

Apply fixtures and external load as discussed earlier. The options in **Goals and Constraints** drop-down are discussed next.

Goals and Constraints

The options displayed under **Goals and Constraints** drop-down are used to set the goals and constraint for the selected component or model that drive the mathematical formulation of the optimization algorithm. You need to select an optimization goal from the **Goals and Constraints** drop-down. The options of **Goals and Constraints** drop-down are discussed next.

The **Goals and Constraints** tool acts like a advisor which helps us to set the goals and constraint for the model. The procedure to use this tool is discussed next.

- Click on the **Goals and Constraints** tool from **Goals and Constraints** drop-down in the **Ribbon**. The **Task List - Topology Study** task pane will be displayed; refer to Figure-5.

Figure-5. Task List Topology Study task pane

- The blue tick in the task pane display the completed action for the particular topology study and the highlighted option is the one on which you need to work.
- Click on the **Set Goals and Constraints** button from the task pane. The **Goals and Constraints** drop-down in **Ribbon** will starts blinking. The blinking of a drop-down or button means, you need to select the required option from the blinking drop-down i.e. **Goals and Constraints** drop-down. If click on the another option from task pane, the button related to selected option will starts blinking and shows the selected parameter of the model on the canvas.
- The selection procedure and working of the goals and constraint of **Goals and Constraints** drop-down are discussed next.

Minimizing Mass with Displacement Constraints

The **Minimize Mass with Displacement Constraints** tool is used to minimize the mass of component while keeping displacement under specified range. The procedure to use this tool is given next.

- Click on the **Goals and Constraints** drop-down from **Ribbon** and click on the **Minimize Mass with Displacement Constraints** button; refer to Figure-6. The **Goals and Constraints PropertyManager** will be displayed; refer to Figure-7.

Figure-6. Goals and Constraints drop-down

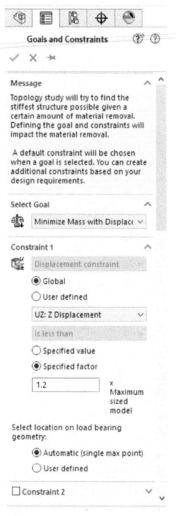

*Figure-7. Goals and Constraints
PropertyManager*

- The **Minimize Mass with Displacement Constraint** option is selected by default in **Select Goal** section. This option is used to create a shape of the selected model which weighs less than the original model without violating the given target for displacement constraint. This option is generally used to reduce the mass of the selected model or component while restricting the displacement under a certain limit.

- The **Displacement Constraint** option is selected by default in **Constraints** drop-down of **Constraint 1** section. This option is used to set the upper limit for the selected displacement component.
- Click in the **Component** drop-down from **Constraint 1** section and select the desired displacement variable.
- Select the **Specified value** radio button from **Constraint 1** section to specify the desired value of selected displacement variable.
- Click in the **Units** drop-down from **Constraint** section and select the desired unit from the list.
- Click in the **Constraint Value** edit box from **Constraint 1** section and enter the desired value of displacement variable if the **Specified value** radio button is selected.
- Select the **Specified factor** radio button from **Constraint 1** section to enter a value of factor to multiply the maximum displacement calculated from a static study.
- Click in the **Constraint value** edit box from **Constrain 1** section and enter the value of factor. The entered value will be multiplied with **Maximum sized model** giving formula for displacement constraints.
- Select the **Automatic (Single max point)** radio button to observe the maximum displacement vertex of the model.
- Select the **User defined** radio button from **Constraint 1** section to select the reference vertex for the displacement constraint. On selecting this radio button, a pink selection box is displayed and activated by default. You need to select a vertex for displacement from the model. The selected vertex will be used as reference.
- Select the **Constraint 2** check box from **Goals and Constraints PropertyManager** and specify the second constraint for the model. The options and parameters of **Constraint 2** check box are similar to **Constraint 1**.
- After specifying the parameters, click on the **OK** button from **Goals and Constraints PropertyManager**. The **Displacement Constraint** will be created and displayed under **Goals and Constraints** node in **Analysis Tree**; refer to Figure-8.

Figure-8. Goals and Constraints node

- Right-click on the **Mesh** button from **Analysis Tree** and click on the **Create Mesh** button from the displayed right-click menu.

- The **Mesh PropertyManager** will be displayed; refer to Figure-10. Adjust the mesh density to **Coarse** by moving the slider to solve the analysis faster.

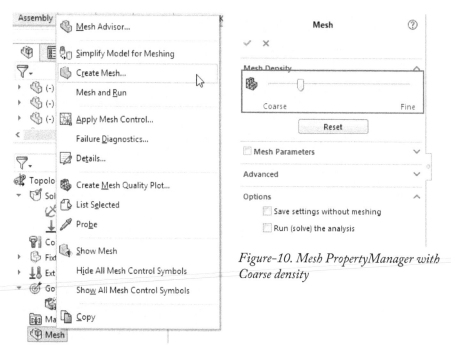

Figure-9. Create Mesh button

Figure-10. Mesh PropertyManager with Coarse density

- After specifying the quality of mesh, click on the **OK** button from **Mesh PropertyManager**. The mesh will be generated and displayed on the model; refer to Figure-11.

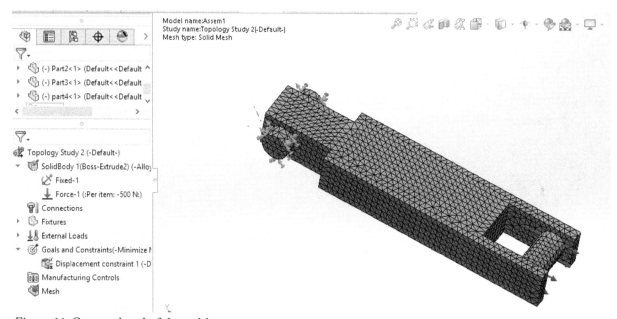

Figure-11. Generated mesh of the model

- Click on the **Run This Study** button from **Ribbon**. The **Topology Study 1** dialog box will be displayed while calculation is running.
- After the completion of calculations, the result will be displayed; refer to Figure-12.

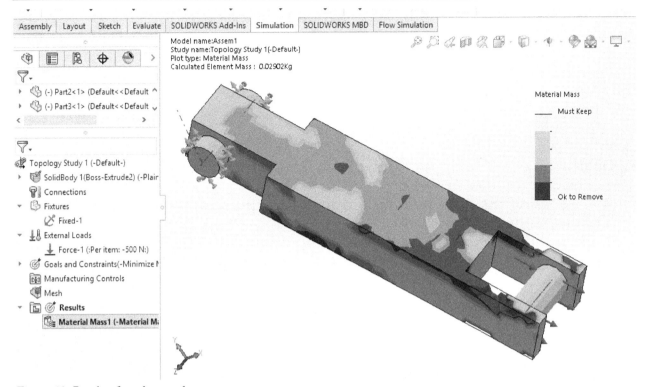

Figure-12. Results of topology study

- On the model, you can see different colors and a chart of **Material Mass**. The dark blue color on the model show the material of the component which can be removed and dark yellow color on the model show the material which must be kept for the strength and stiffness of the particular component.
- While zooming the model, you can see some spots on the model which are removed by the software, this means the following material is not necessary for the component and this removed material will not effect the strength and stiffness of the component; refer to Figure-13.

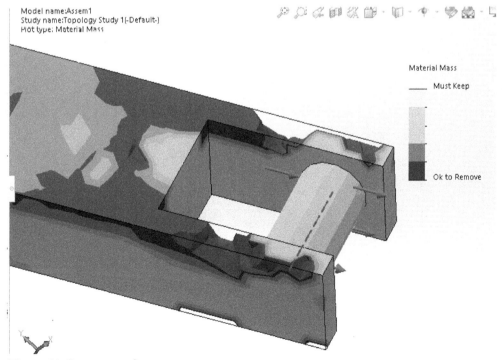

Figure-13. Component after zoom

Best Stiffness to Weight Ratio (default)

The **Best Stiffness to Weight Ratio (default)** option creates a shape of the selected model or component with the largest stiffness considering the given amount of mass that can be removed. On selecting this option, the calculation seeks to minimize the compliance of the model which is a measure of flexibility. The compliance used in this definition implies the sum of strain energies of all elements. The procedure to use this option is discussed next.

- Open the desired model and activate the **Simulation** tab. Apply load and fixtures as discussed earlier in this book.
- We are using the same model and applying the same conditions as used in discussing the previous option.
- Click in the **Goals and Constraints** drop-down from **Ribbon** in the **Simulation** tab and click on the **Best Stiffness to Weight Ratio (default)** button; refer to Figure-14. The **Goals and Constraints PropertyManager** will be displayed; refer to Figure-15.

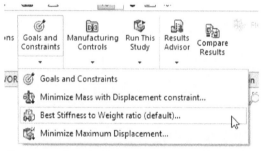

Figure-14. Best Stiffness to weight ratio (default)

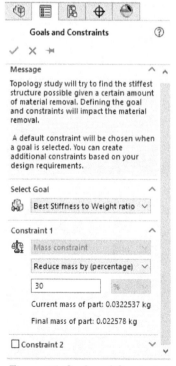

Figure-15. Goals and Constraints PropertyManager

- The **Best Stiffness to Weight Ratio (default)** option is selected by default in **Select Goal** section.
- The **Mass Constraint** option is selected by default in **Constraint** box from **Constraint 1** section. The **Mass Constraint** option is used to set the targeted mass of the part up to which the mass will be reduced after calculation.
- Click in the **Component** drop-down from **Constrain 1** section and select **Reduce mass by (absolute value)** option to enter the value of mass which is to be removed from part.

Or

- Click on the **Component** drop-down from **Constrain 1** section and select **Reduce Mass by (Percentage)** option to enter the targeted percentage of mass reduction.
- Click in the **Constrain Value** edit box and enter the desired value of mass constraint. Note that the reduction of mass from the original shape of model will be attempted by optimization through an iterative process.
- Click on the **Constrain 2** check box to specify the parameters for another constraint. The options similar to **Constraint 1** will be displayed.
- After specifying the parameters, click on the **OK** button from **Goals and Constraints PropertyManager**. The **Mass Constraint** option will be added in **Analysis Tree** below **Goals and Constraints** node; refer to Figure-16.

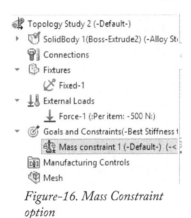

Figure-16. Mass Constraint
option

- Create the mesh of current model up to required intensity of mesh. The procedure of creating mesh was discussed earlier.
- Click on the **Run This Study** button from **Ribbon** in the **Simulation** tab. The **Topology Study 2** dialog box will be displayed.
- After calculation are complete, results will be displayed in **Results** node of **Analysis Tree**; refer to Figure-17.

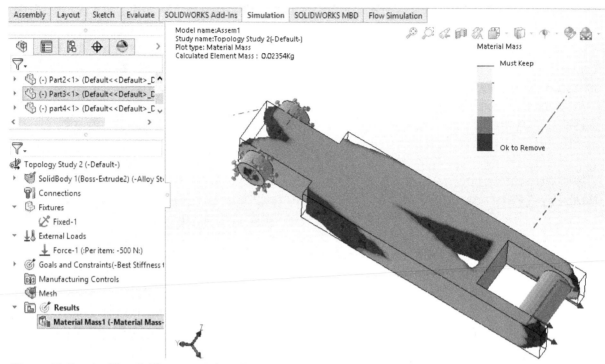

Figure-17. Result of Best Stiffness to weight ratio

- On the canvas screen, the **Material Mass** legend will be displayed on the basis of optimization of final shape.
- Some parts or spots of the final shape will disappear which means the spots can be removed from the model. The shadow of disappeared part is highlighted in blue color. For reducing mass without altering the strength and stiffness of model, one can remove the blue highlighted spots or parts from the model.

Minimize Maximum Displacement

The **Minimize Maximum Displacement** option creates a shape that minimize the maximum displacement on a single node. In this study, we provide a percentage of material to be removed from a component. The software optimize model to stiffest design that weighs less than the initial design of component. It also minimize the maximum observed displacement. The procedure to use this option is discussed next.

- Open the desired model in **Simulation** environment and apply load & fixtures as discussed earlier in this book. We are using the same model and applying the same conditions as discussed in previous sections.
- Click on the **Goals and Constraints** drop-down from **Ribbon** on the **Simulation** tab and select the **Minimize Maximum Displacement** tool; refer to Figure-18. The **Goals and Constraints PropertyManager** will be displayed; refer to Figure-19.

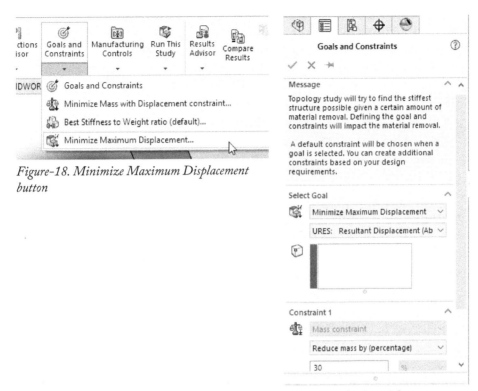

Figure-18. Minimize Maximum Displacement button

Figure-19. Goals and Constraints PropertyManager

- The **Minimize Maximum Displacement** option is selected by default in **Goals and Constraints** drop-down in the **Select Goal** section.
- Click in the **Component** drop-down from **Select Goal** section and select the desired displacement variable for optimization.
- Click in the **Vertex for Displacement** selection box and select the desired vertices for minimizing the maximum displacement; refer to Figure-20.

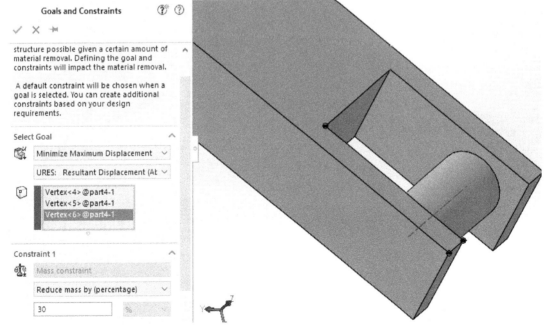

Figure-20. Selecting vertex

- The **Mass Constraint** button is selected by default in **Constraints** drop-down of **Constraint 1** section. The **Mass Constraint** option is used to set the targeted mass of the part that will be reduced during optimization.
- Click on the **Component** drop-down from **Constrain 1** section and select **Reduce mass by (absolute value)** option to enter the value of mass which is to be removed from part.
- Click on the **Component** drop-down from **Constrain 1** section and select **Reduce Mass by (Percentage)** option to enter the targeted percentage of mass reduction.
- Click in the **Constrain Value** edit box and enter the desired value of mass constraint
- After specifying the parameters, click on the **OK** button from **Goals and Constraints PropertyManager**. The **Mass Constraint** option will be added in **Goals and Constraints** node in **Analysis Tree**.
- Create the mesh of current model as required intensity of mesh. The procedure of creating mesh was discussed earlier.
- Click on the **Run This Study** button from **Ribbon** in the **Simulation** tab. The **Topology Study** dialog box will be displayed.
- After calculations, results will be displayed in **Results** node of **Analysis Tree**; refer to Figure-21.

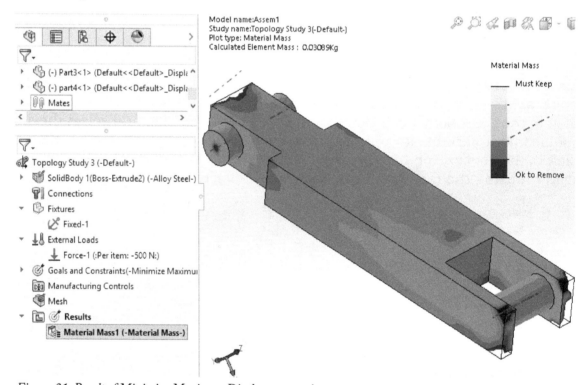

Figure-21. Result of Minimize Maximum Displacement option

- Check the vertices selected for the current model by zooming on the respective locations; refer to Figure-22. You will find the displacement of the selected vertices is minimum as compared to the similar corner of other sides.

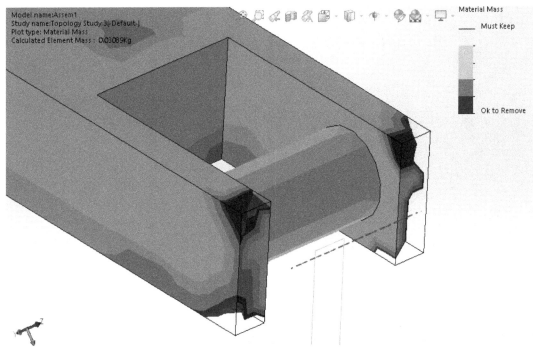

Figure-22. Zooming the vertex

- After performing topology study, you can add more mass or displacement constraints by selecting the **Add Displacement Constraint** or **Add Mass Constraint** button from **Goals and Constraints** drop-down; refer to Figure-23.

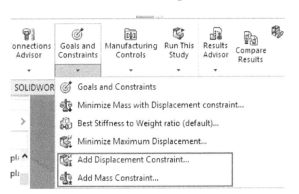

Figure-23. Adding Displacement or Mass constraint button

Section Clipping

The **Section Clipping** tool is used to create section views for displayed result. You can define three section using this tool like Planer sections, Cylindrical sections, and spherical sections. The procedure to use this is discussed next.

- Right-click on the **Material Mass** result form **Analysis Tree**, the right-click shortcut menu will be displayed; refer to Figure-24.

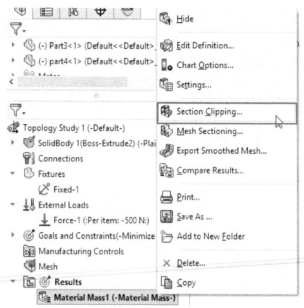

Figure-24. Section Clipping tool

- Click on the **Section Clipping** tool from the displayed menu, the **Section PropertyManager** will be displayed; refer to Figure-25

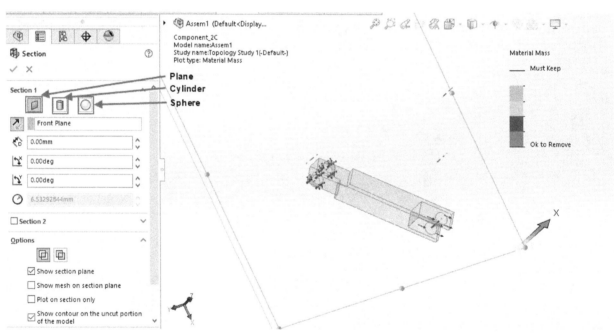

Figure-25. Section PropertyManager

- Click on the **Plane** button for planer sections using any reference plane. Select a plane as per requirement.
- Click on the **Cylindrical** button from cylindrical sections using a reference axis or plane. Select a reference axis or a plane so that section view of the part or component will be created using cylindrical protrusion of the selected reference.
- Click on the **Spherical** button for spherical sections using a reference point, a 3D point, or a vertex. You need to select a point or a vertex and specify the parameters like radius, distance from reference and so on to create spherical sections.
- In our case, we have selected the **Plane** button. The **Front Plane** is selected by default in **Reference entity** selection box. The **Front Plane** will be displayed near model; refer to Figure-26

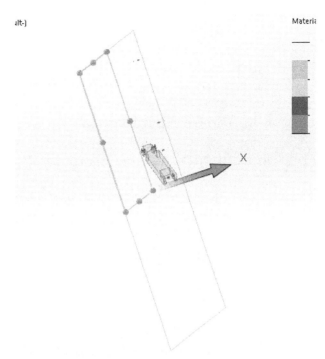

Figure-26. Front Plane near model

- You can see the red arrow at the centre of plane. Grab the arrow and move this towards model. The plane starts moving towards model or towards the direction of arrow head. When the plane cuts the model at desired location, you can see the preview of section view; refer to Figure-27.

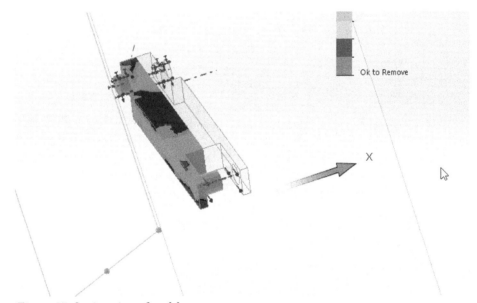

Figure-27. Section view of model

- You can also do the above process by entering the value of distance in **Distance** edit box from **Section** section.
- Click on the **Reverse clipping Direction** button from **Section 1** section to invert the section view of model.
- Click in the **Rotation X** edit box from **Section 1** section and enter the desired value in degree to rotate the selected plane for section view along X direction.

- Click in the **Rotation Y** edit box from **Section 1** section enter the desired value in degree to rotate the above selected plane for section along in Y direction.
- The **Radius** edit box is displayed when you had selected the **Cylinder** or **Spherical** button. Specify the desired value of radius as required.
- Select the **Section 2** check box to use additional section plane. The procedure of creating the section view in **Section 2** is same as discussed above.
- Click on the **Intersection** button from **Options** section to display the intersection area of all sections.
- Click on the **Union** button from **Options** section to display the union area of all sections.
- Select the **Show Section Plane** check box from **Options** section to view the section plane in the canvas area or graphics area. Clear the check box to hide the plane from canvas area.
- Select the **Show mesh on section plane** check box from **Options** section to view generated mesh on the model; refer to Figure-28. Clear the check box to hide mesh.

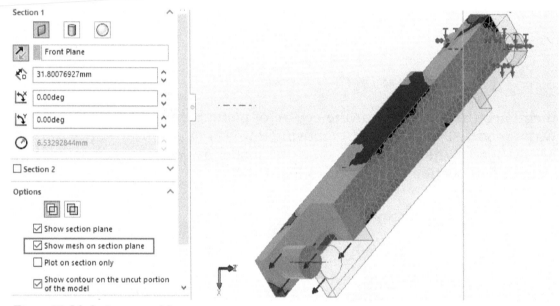

Figure-28. Mesh view on model

- Select the **Plot on section only** check box from **Options** section to view the result contour on the selected section only. This will show only the selected region which comes across section plane; refer to Figure-29. Clear this check box to disable this function.

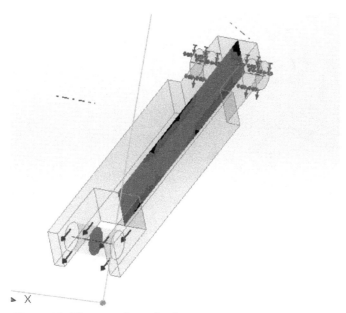

Figure-29. Plot on section only view

- Select the **Show contour on the uncut portion of the model** check box from **Options** section to view the result contours of uncut portions of model. Clear this check box to see the uncut portion of the model in shaded view; refer to Figure-30. This option could only be activated when **Plot on section only** check box is cleared.

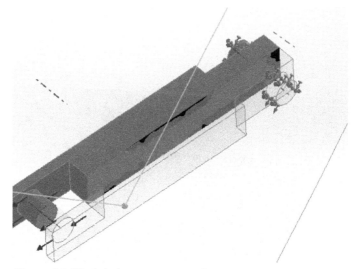

Figure-30. Shaded view on uncut portion

- Select the **Explode after clipping** check box from **Options** section to explode the model after creating the section plot. Clear this check box to create the section plot on the exploded view directly.
- Click on the **Reset** button from **Options** section to reset all the settings.
- After specifying the parameters, click on the **OK** button from **Section PropertyManager**. The selected changes will be applied to the view.

Mesh Sectioning

The **Mesh Sectioning** tool is used to view mesh elements inside the component. The procedure to use this tool is discussed next.

- Right-click on the **Material Mass** result, the right-click menu will be displayed.

Figure-31. Mesh Sectioning button

- Click on the **Mesh Sectioning** button from the displayed menu; refer to Figure-31. The **Mesh Section PropertyManager** will be displayed; refer to Figure-32, along with sectioned model and generated mesh on section plane.

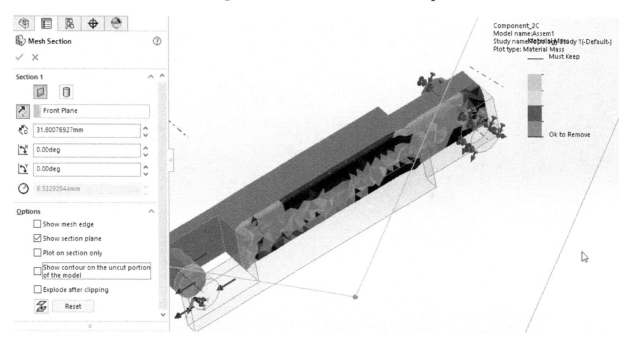

Figure-32. Mesh Section PropertyManager

- The parameters and options of **Mesh Section PropertyManager** are similar to **Section PropertyManager**.
- All other option works in the same way as discussed earlier.
- Select the **Show mesh edge** check box from **Options** section to view the generated mesh on the model. Clear this to hide the mesh.
- After specifying the parameters, click on the **OK** button from **Mesh PropertyManager.**

Export Smoothed Mesh

The **Export Smoothed Mesh** tool is used to export the current smoothed mesh data of the optimized model shape as a new geometry. The procedure to use this option is discussed next.

- Right-click on the **Material Mass** result from **Analysis Tree**. The Right-click menu will be displayed; refer to Figure-33.

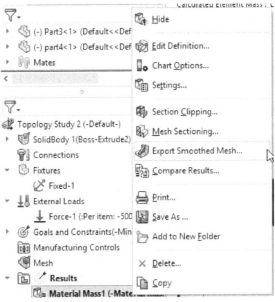

Figure-33. Export Smoothed Mesh button

- Click on the **Export Smoothed Mesh** button from the displayed menu. The **Export Smoothed Mesh PropertyManager** will be displayed; refer to Figure-34.
- The **Smoothened** mesh can be saved in active configuration, or in the form of a new configuration, or a new part.

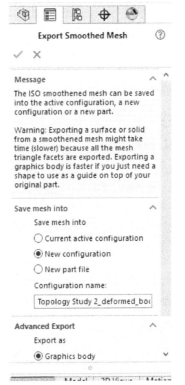

Figure-34. Export Smoothed Mesh PropertyManager

- Select the **Current active configuration** radio button from **Save mesh into** section to save the smoothened data in current configuration.
- Select the **New Configuration** radio button from **Save mesh into** section to save the smoothened mesh data in a new configuration. The configuration name is displayed as default in **Configuration Name** edit box. If you want to change the name, click in the **Configuration Name** edit box and enter the desired name.
- Select the **New part file** radio button from **Save mesh into** section to save the smoothened mesh data in a new document. Specify the name and location of part file to be created.
- Select the **Graphics body** radio button from **Advanced Export** section to export the smoothened mesh data in graphics body. Software exports the smoothened mesh data in a lightweight, boundary geometry representation format. This option is also used to import the graphics body document into assembly document, so that it will be accessible as a blueprint to help modify the geometry of original document.
- Select the **Solid body** radio button from **Advanced Export** section to export the smoothened mesh data as solid body. The format of exported file will be **.sldprt**. This option is generally used in case of 3D printing operations. This option take longer processing or computational time to process the information for export data correlate to other options.
- Select the **Surface Body** radio button from **Advanced Export** section to export only the surface geometry of the smoothened mesh. The format of exported file will be **.STL**.
- After setting the required parameters and click on the **OK** button from the **Export Smoothened Mesh PropertyManager**. You will be redirected to default screen as per entered data.

MANUFACTURING CONTROLS

The optimization process of SolidWorks creates a layout of material that satisfies the optimization goals. Sometimes, the design is to be created using standard manufacturing techniques like forging or casting. In these case, the tools of **Manufacturing Controls** are used to preserve the important manufacturing locations so that, the engineer are able to form the component using a mold or with the use machines. The procedure to use the tools of **Manufacturing Controls** drop-down are discussed next.

Manufacturing Controls

The **Manufacturing Controls** tool works as an advisor to deal with the tools related to Manufacturing Controls. To use **Manufacturing Controls** tool, you need to run the study after applying the desired goals and constraints. This process is vital because after this we are able to understand the component demand and requirement in practical use.

- Click on the **Manufacturing Controls** tool from **Manufacturing Controls** drop-down in **Ribbon**. The **Topology Study** pane will be displayed at the left in the screen; refer to Figure-35.

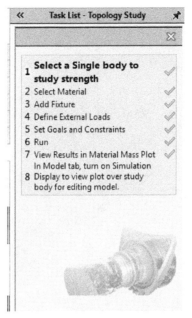

Figure-35. Topology Study Task Pane

- The Blue tick in front of the particular option or tool refers to the completed task that is completed earlier. If blue tick is unavailable in the pane for any task then it means something is incomplete. Complete the missing steps as discussed earlier.

Add Preserved Region

The **Add Preserved Region** tool is used to add a preserved region of your model that will bot be modified during topology optimization. In this, you are able to stiffen the selected regions of you model that are contacting other parts, like the regions used to fix the model, regions used to make connections with other parts. These selected region do not participate in topology study. The procedure to use this is discussed next.

- Click on the **Add Preserved Region** tool from **Manufacturing Controls** drop-down in the **Ribbon;** refer to Figure-36. The **Preserved Region PropertyManager** will be displayed; refer to Figure-37.

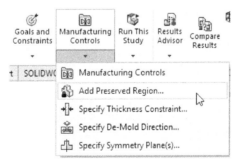

Figure-36. Add Preserved Region button

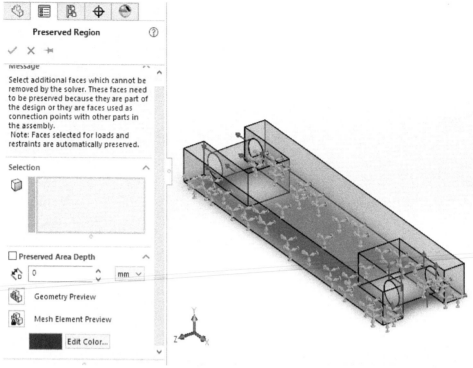

Figure-37. Preserved Region PropertyManager

- The selection box of **Preserved Region PropertyManager** is activated by default. You need to select those face which you want to be preserved after optimization process. These faces may be the connection point with other component of assembly. Do not select the faces which are selected for loads and restraints because these kind of faces were automatically selected.
- Select the faces as desired from model in canvas screen; refer to Figure-38. The selected faces will be highlighted and displayed in **Preserved Region PropertyManager**.

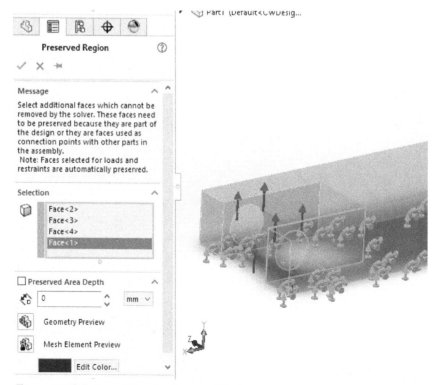

Figure-38. Selection of faces for Preserved Region

- Select the **Preserved Area Depth** check box from **Preserved Region PropertyManager** to set the depth of region that will remain unchanged after study. The related parameters will be displayed.
- Click in the **Preserved Area Depth** edit box and enter the desired value of depth.
- Click on the **Unit** drop-down and select the desired unit for entering the value of area depth.
- Click on the **Geometry Preview** button from **Preserved Area Depth** section to view the geometry of selected faces for preserved region; refer to Figure-39.

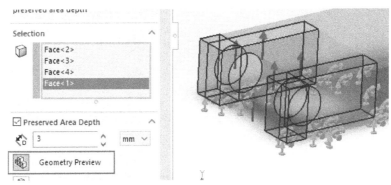

Figure-39. Geometry Preview of preserved region

- Select the **Mesh Element Preview** button from **Preserved Area Depth** section to view the mesh of selected preserved area. The mesh will be displayed in default color; refer to Figure-40.

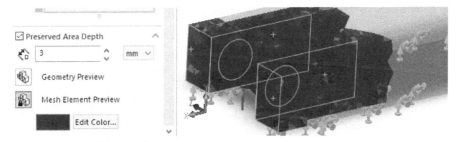

Figure-40. Mesh Element Preview

- If you want to change the color of displayed mesh, click on the **Edit Color** button and select the desired color.
- After specifying the parameters, click on the **OK** button from **Preserved Region PropertyManager**. You will be redirected to Simulation screen.
- Click on the **Run This Study** button from **Ribbon**, the optimization process will be started and results will be displayed on completion of optimization process with preserved regions; refer to Figure-41.

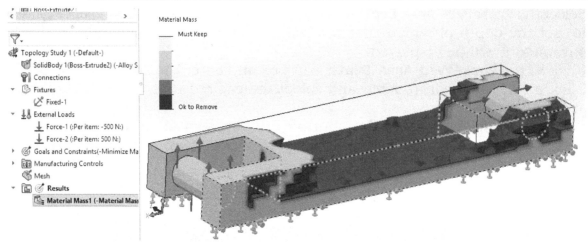

Figure-41. Result of preserved region

Specify Thickness Constraint

The **Specific Thickness Constraint** tool is used to apply size restriction to a topology study that prohibits the creation of very thick or thin regions that may be difficult to manufacture practically. The procedure to use this tool is discussed next.

*   Click on **Specify Thickness Constraint** button from **Manufacturing Controls** drop-down; refer to Figure-42. The **Thickness Control PropertyManager** will be displayed; refer to Figure-43.

Figure-42. Specify Thickness Constraint button

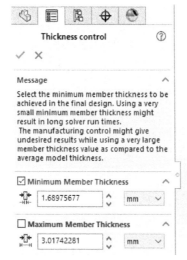

Figure-43. Thickness control PropertyManager

Try to avoid specifying the small minimum number thickness because that might take long solver time for optimization. Similarly, the entered large member thickness value may give undesired results as compared to average model thickness.

- Click in the **Minimum Member Thickness** edit box and enter the desired value of an optimized substructure. The entered value should be greater than two to three times the average element size.
- Click in the **Maximum Member Thickness** edit box and enter the desired maximum value of optimized substructure.
- After specifying the parameters, click on the **OK** button from **Thickness Control PropertyManager**.
- Click on the **Run This Study** button from **Ribbon**, the optimization process will be started and result will be displayed after calculations; refer to Figure-44.

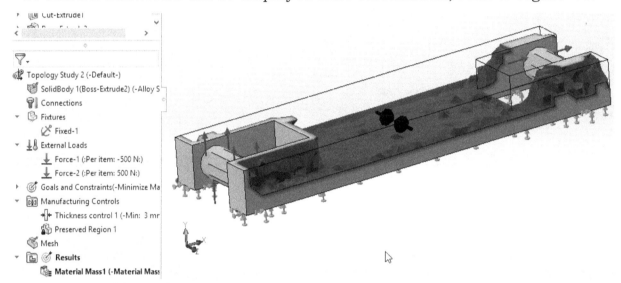

Figure-44. Result of Thickness Control

As you can see in the above picture, we had applied the **Preserved Region** tool on some regions of model and then the **Thickness Control** tool to limit the maximum and minimum value for material removal.

Specify De-mold Direction

The **Specify De-mold Direction** tool is used to add de-mold control to ensure that the optimized design is able to manufacture and can be extracted from a mold. The procedure to use this too is discussed next.

- Click on the **Specify De-mold Direction** button from **Manufacturing Controls** drop-down; refer to Figure-45. The **De-mold Control PropertyManger** will be displayed; refer to Figure-46.

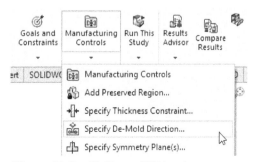

Figure-45. Specify De-mold Direction

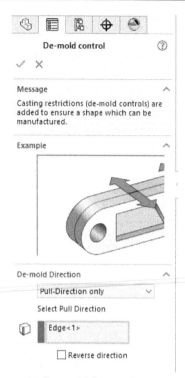

Figure-46. De-mold Control PropertyManager

- Click on the **Select Direction** drop-down from **De-mold Control PropertyManager** and select the desired option.
- The pink selection box is activated by default. You need to select the plane about which you want to apply the De-mold control tool.
- Select the **Determine central mid plane automatically** check box to select the position of mid plane which is optimized automatically if you have selected the **Mid-Plane** option from the **De-mold Direction** drop-down.
- The **Select edge to define the direction** selection box is displayed and activated by default. You need to click on the desired edge from model to define mold opening direction.
- After specifying the parameters, click on the **OK** button from **De-mold control PropertyManager**. You will be redirected to Simulation workplace.
- Click on the **Run This Study** button from **Ribbon**, the optimization process will start and result will be displayed after calculations; refer to Figure-47.

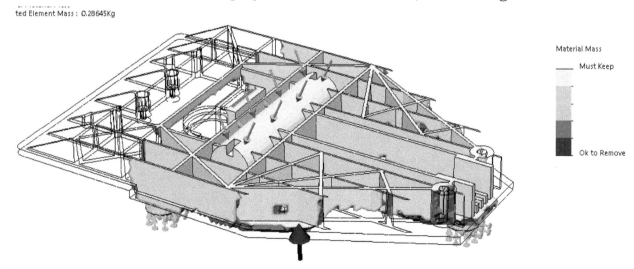

Figure-47. De-mold Control result

Specify Symmetry Plane(s)

The **Symmetry Plane** tool is used to force the optimized design to be in symmetry along a selected plane. You can also enforce a half, quarter, or one-eighth planer symmetry for an optimized design. The procedure to use this tool is discussed next.

- Click on the **Specify Symmetry Plane(s)** tool from **Manufacturing Controls** drop-down in **Ribbon**. The **Symmetry Control PropertyManager** will be displayed; refer to Figure-48.

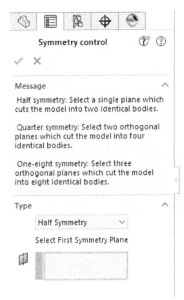

Figure-48. Symmetry Control PropertyManager

- Select the **Half Symmetry** button from **Type** drop-down to select a single plane which give the optimization result identical around the selected plane on a model.
- The **Select First Symmetry Plane** is activated by default. You need to select the desired plane.
- Select the **Quarter Symmetry** button from **Type** drop-down to select two orthogonal plane which identically optimize the model body into 4 identical portions.
- The **Select Second Symmetry Plane** is activated by default. You need to select the second plane.
- Select the **One-Eight Symmetry** button from **Type** drop-down to select three orthogonal planes which optimize the model identical into eight portions.
- The **Select Third Symmetry Plane** is activated by default. You need to select the third plane.
- After specifying the required parameters, click on the **OK** button from **Symmetry Control PropertyManager**.
- Click on the **Run This Study** button from **Ribbon**, the optimization process will be started and result will be displayed after calculations; refer to Figure-49.

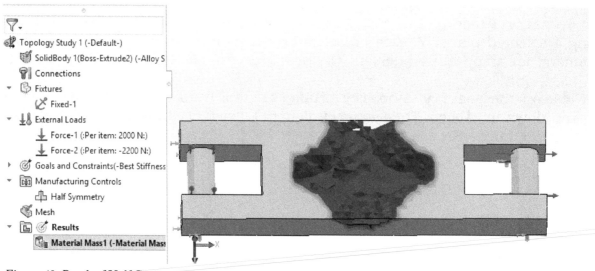

Figure-49. Result of Half Geometry

PRACTICAL

Perform the Topology Study on the wrench to get optimum shape with minimum mass under the specified load condition; refer to Figure-50.

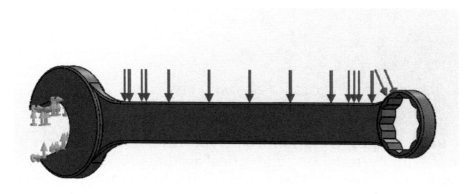

Figure-50. Practical 1 Topology Study

- Open the wrench model from resource folder of SolidWorks Simulation 2020. The wrench model will be displayed in SolidWorks.
- Click on the **New Study** tool from **Ribbon** of **Simulation** tab, the **Study PropertyManger** will be displayed; refer to Figure-51.

*Figure-51. Study PropertyMan-
ager Topology Study*

• Click in the **Study Name** edit box from **Name** section and enter the name as **Wrench topology study**. Click on the **OK** button from **Study PropertyManager**. The **Simulation** tab will be displayed; refer to Figure-52.

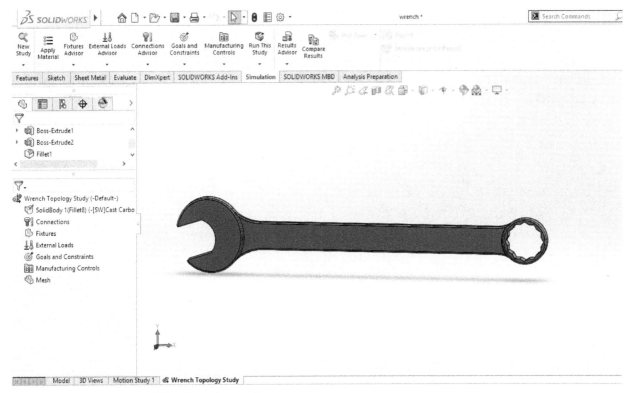

Figure-52. Simulation tab of Wrench topology study

Applying Material

- Click on the **Apply Material** button from **Ribbon** of **Simulation** tab. The **Material** dialog box will be displayed; refer to Figure-53.

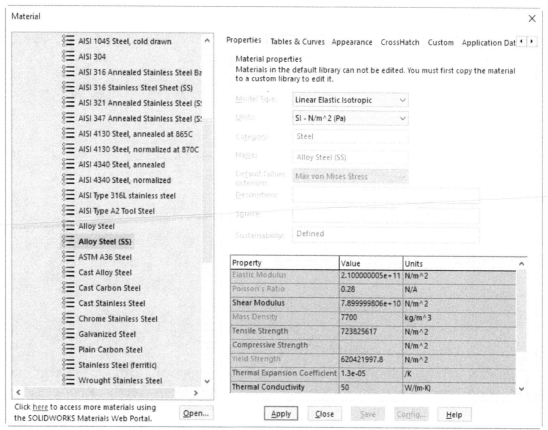

Figure-53. Material selection for wrench

- Select the **Alloy Steel (SS)** from material list and click on the **Apply** button from **Material** dialog box. The selected material will be applied to your model.
- Click on the **Close** button from **Material** dialog box. The selected material will be applied; refer to Figure-54.

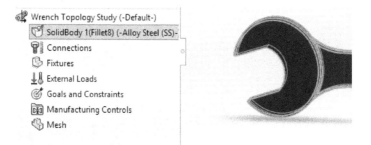

Figure-54. Mentioned name in analysis tree

Applying Fixture

- Click on the **Fixed Geometry** button from **Fixture Advisor** drop-down of **Ribbon**. The **Fixture PropertyManager** will be displayed; refer to Figure-55.

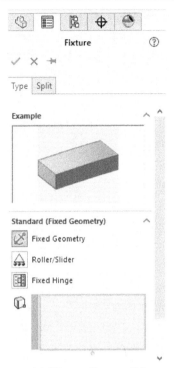

Figure-55. Fixture PropertyManager 1

- The selection box is active by default. Select the faces of wrench as displayed in Figure-56.

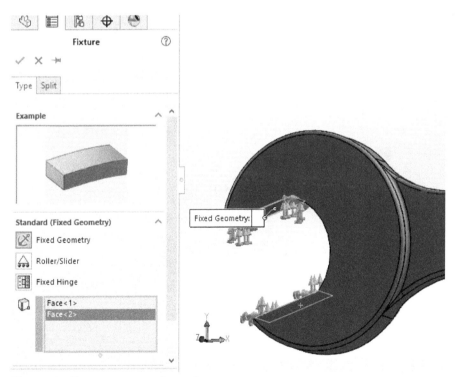

Figure-56. Fixing geometry of wrench

- After selecting the faces, click on the **OK** button from **Fixture PropertyManager**, the fixed geometry button will be displayed in **Analysis** tree.

Applying Force

- Click on the **Force** button from **External Loads Advisor** drop-down of **Ribbon**, the **Force/Torque PropertyManager** will be displayed; refer to Figure-57.

Figure-57. Force Torque PropertyManager

- The selection box is active by default. You need to select the faces as displayed in Figure-58 for applying force. The pink color arrow indicates the direction applied force.

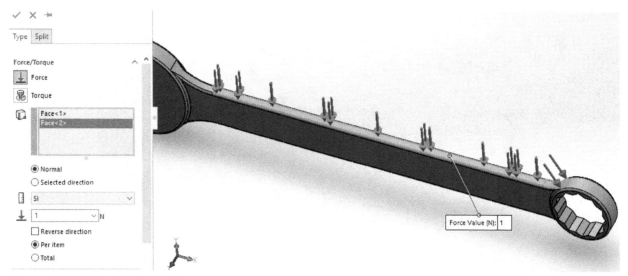

Figure-58. Selecting face for applying force

- Click in the **Force Value** edit box from the **Force/Torque PropertyManager** and enter the value as **300 N**.
- After specifying the parameters, click on the **OK** button from the **Force/Torque PropertyManager**. The **Force-1** will be displayed in **Analysis** tree.

Applying Goals and Constraints

- Click on the **Minimize Mass with Displacement constraint** button from **Goals and Constraints** drop-down of **Ribbon**. The **Goals and Constraints PropertyManager** will be displayed; refer to Figure-59

Figure-59. Goals and Constraints PropertyManager

- Click on the **OK** button from **Goals and Constraints PropertyManager**. The **Displacement Constraint 1** will be displayed in **Analysis tree**; refer to Figure-60.

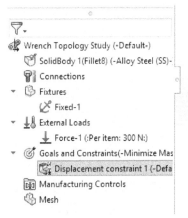

Figure-60. Displacement Constraint in analysis tree

Adding Preserved Region

- Click on the **Add Preserved Region** button from **Manufacturing Controls** drop-down on **Ribbon**. The **Preserved Region PropertyManager** will be displayed; refer to Figure-61.

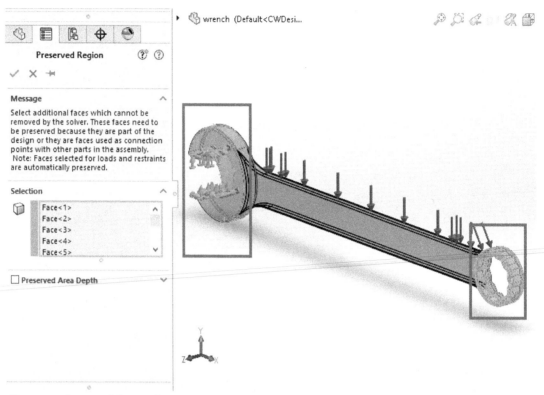

Figure-61. Preserved Region for wrench

- The selection box of **Preserved Region PropertyManager** is activated by default. You need to select all the faces highlighted in a box in the above figure. You can use box selection to select these faces.

Note- Do not select the faces on which you had applied force. Because these faces are preserved region by default.

- After selecting all the faces from model, click on the **OK** button from **Preserved Region PropertyManager**. The **Preserved Region** feature will be added in **Analysis Tree** under **Manufacturing Controls**.

Creating Mesh

We had applied all the required tools and parameters for creating a topology study. Before generating the results, we need to create a mesh. If you do not create mesh at this point then the software automatically create the mesh with default parameters. But software may create a fine mesh which will take more computational time for optimization. So, we are creating the mesh of wrench by our own. The procedure to create mesh is discussed next.

- Click on the **Create Mesh** button from **Run This Study** drop-down of **Ribbon**; refer to Figure-62. The **Mesh PropertyManager** will be displayed; refer to Figure-63.

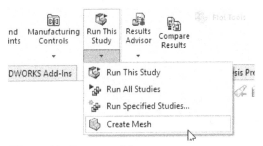

Figure-62. Create mesh button

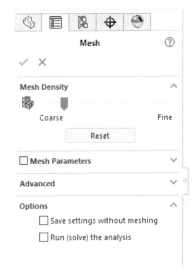

Figure-63. Mesh setting for wrench

- Set the mesh slider as per the previous figure and click on the **OK** button from **Mesh PropertyManager**. The mesh of model will be generated and displayed; refer to Figure-64.

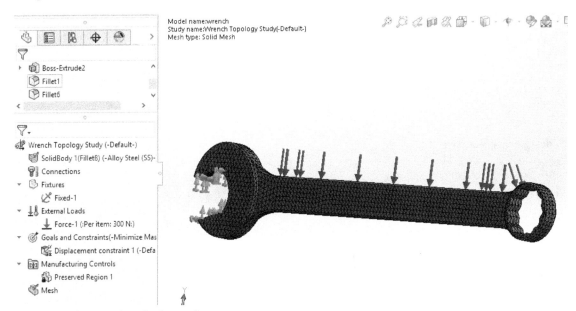

Figure-64. Generated mesh of wrench

Results

Till now, we have applied fixtures and force to the wrench model. To deal with fast computing, we had also created coarse mesh of the model. Now its time for results.

- Click on the **Run This Study** button from **Ribbon** to start the process of optimization. The **Wrench Topology Study** dialog box will be displayed with iteration calculations going on; refer to Figure-65. Click on the **View Convergence Data and Graphs** button to view the respective graphs.

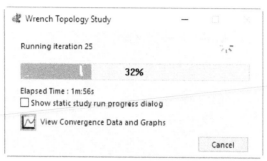

Figure-65. Wrench Topology Study dialog box

- Result of the following study will be displayed on completion of optimization; refer to Figure-66.

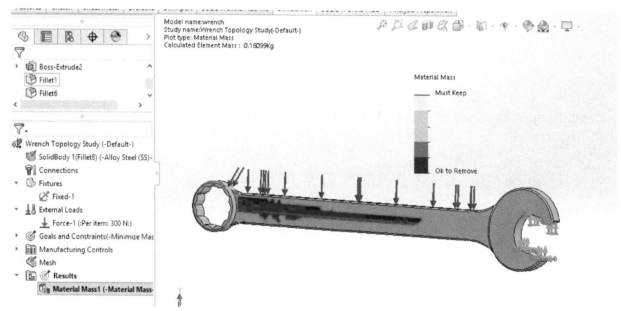

Figure-66. Result of wrench topology study

- Now, you can remove the not required material by using modeling tools of SolidWorks.

FOR STUDENT NOTES

FOR STUDENT NOTES

Chapter 15

Fundamentals of FEA

Topics Covered

The major topics covered in this chapter are:

- *Introduction to FEA.*
- *General Description of FEA*
- *Solving a problem with FEA*
- *FEA V/s Classical Methods*
- *FEA V/s Finite Different Method*
- *Need for Studying FEM*
- *Refining Mesh*
- *Higher order element V/s Refined Mesh*

INTRODUCTION

The finite element analysis is a numerical technique. In this method all the complexities of the problems, like varying shape, boundary conditions and loads are maintained as they are but the solutions obtained are approximate. Because of its diversity and flexibility as an analysis tool, it is receiving much attention in engineering. The fast improvements in computer hardware technology and slashing of cost of computers have boosted this method, since the computer is the basic need for the application of this method. A number of popular brand of finite element analysis packages are now available commercially. Some of the popular packages are SolidWorks Simulation, NASTRAN, NISA, and ANSYS. Using these packages one can analyze several complex structures.

The finite element analysis originated as a method of stress analysis in the design of aircrafts. It started as an extension of matrix method of structural analysis. Today this method is used not only for the analysis in solid mechanics, but even in the analysis of fluid flow, heat transfer,electric and magnetic fields and many others. Civil engineers use this method extensively for the analysis of beams, space frames, plates, shells, folded plates, foundations, rock mechanics problems and seepage analysis of fluid through porous media. Both static and dynamic problems can be handled by finite element analysis. This method is used extensively for the analysis and design of ships, aircraft, space crafts, electric motors, and heat engines.

GENERAL DESCRIPTION OF FEM

In engineering problems, there are some basic unknowns. If they are found, the behavior of the entire structure can be predicted. The basic unknowns or the Field variables which are encountered in the engineering problems are displacements in solid mechanics, velocities in fluid mechanics, electric and magnetic potentials in electrical engineering and temperatures in heat flow problems.

In a continuum, these unknowns are infinite. The finite element procedure reduces such unknowns to a finite number by dividing the solution region into small parts called elements and by expressing the unknown field variables in terms of assumed approximating functions (Interpolating functions/Shape functions) within each element. The approximating functions are defined in terms of field variables of specified points called nodes or nodal points. Thus in the finite element analysis, the unknowns are the field variables of the nodal points. Once these are found the field variables at any point can be found by using interpolation functions.

After selecting elements and nodal unknowns next step in finite element analysis is to assemble element properties for each element. For example, in solid mechanics, we need to find the force-displacement i.e. stiffness characteristics of each individual element. Mathematically this relationship is of the form

$$[k]_e \{\delta\}_e = \{F\}_e$$

where $[k]_e$ is element stiffness matrix, $\{\delta\}_e$ is nodal displacement vector of the element and $\{F\}_e$ is nodal force vector. The element of stiffness matrix k_{ij} represent the force in coordinate direction 'i' due to a unit displacement in coordinate direction 'j'. Four methods are available for formulating these element properties viz. **Direct approach**, **Variational approach**, **Weighted Residual approach**, and **Energy balance approach**. Any one of these methods can be used for assembling element properties. In solid mechanics variational approach is commonly employed to assemble stiffness matrix and nodal force vector (consistent loads).

Element properties are used to assemble global properties/structure properties to get system equations $[k]\{\delta\} = \{F\}$. Then the boundary conditions are imposed. The solution of these simultaneous equations give the nodal unknowns. Using these nodal values, additional calculations are made to get the required values e.g. stresses, strains, moments, etc. in solid mechanics problems.

Thus the various steps involved in the finite element analysis are:

(i)   Select suitable field variables and the elements.
(ii)   Discretize the continua.
(iii)   Select interpolation functions.
(iv)   Find the element properties.
(v)   Assemble element properties to get global properties.
(vi)   Impose the boundary conditions.
(vii)   Solve the system equations to get the nodal unknowns.
(viii) Make the additional calculations to get the required values.

A BRIEF EXPLANATION OF FEA FOR A STRESS ANALYSIS PROBLEM

The steps involved in finite element analysis are clarified by taking the stress analysis of a tension strip with fillets (refer Figure-1). In this problem stress concentration is to be studies in the fillet zone. Since the problem is having symmetry about both x and y axes, only one quarter of the tension strip may be considered as shown in Figure-2. About the symmetric axes, transverse displacements of all nodes are to be made zero. The various steps involved in the finite element analysis of this problem are discussed below:

Step 1: Four noded isoparametric element are selected for the analysis (However note that 8 noded isoparametric element is ideal for this analysis). The four noded isoparametric element can take quadrilateral shape also as required for elements 12, 15, 18, etc. As there is no bending of strip, only displacement continuity is to be ensured but not the slope continuity. Hence, displacements of nodes in x and y directions are taken as basic unknowns in the problem.

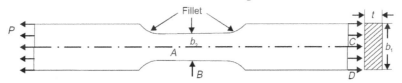

Figure-1. Tension Strip with Fillet

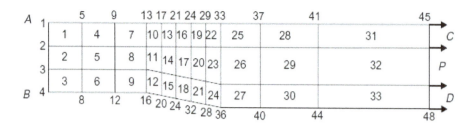

Figure-2. Discretization of quarter of tension strip

Step 2: The portion to be analyzed is to be discretized. Figure-2 shows discretized portion. For this 33 elements have been used. There are 48 nodes. At each node unknowns are x and y components of displacements. Hence in this problem total unknowns (displacements) to be determined are 48 × 2 = 96.

Step 3: The displacement of any point inside the element is approximated by suitable functions in terms of the nodal displacements of the element. For the typical element, displacements at P are

$$u = \Sigma N_i u_i = N_1 u_1 + N_2 u_2 + N_3 u_3 + N_4 u_4$$
and
$$v = \Sigma N_i v_i = N_1 v_1 + N_2 v_2 + N_3 v_3 + N_4 v_4 \ ...(1.2)$$

The approximating functions N_i are called shape functions or interpolation functions. Usually, they are derived using polynomials.

Step 4: Now the stiffness characters and consistent loads are to be found for each element. There are four nodes and at each node degree of freedom is 2. Hence degree of freedom in each element is 4 × 2 = 8. The relationship between the nodal displacements and nodal forces is called element stiffness characteristics. It is of the form

$[k]_e \{\delta\}_e = \{F\}_e$, as explained earlier.

For the element under consideration, k_e is 8 × 8 matrix and δ_e and F_e are vectors of 8 values. In solid mechanics, element stiffness matrix is assembled using variational approach i.e. by minimizing potential energy. If the load is acting in the body of element or on the surface of element, its equivalent at nodal points are to be found using variational approach, so that right hand side of the above expression is assembled. This process is called finding consistent loads.

Step 5: The structure is having 48 × 2 = 96 displacement and load vector components to be determined.

Hence global stiffness equation is of the form

[k]   {δ}   =   {F}
96 × 96   96 × 1   96 × 1

Each element stiffness matrix is to be placed in the global stiffness matrix appropriately. This process is called assembling global stiffness matrix. In this problem force vector F is zero at all nodes except at nodes 45, 46, 47 and 48 in x direction. For the given loading nodal equivalent forces are found and the force vector F is assembled.

Step 6: In this problem, due to symmetry transverse displacements along AB and BC are zero. The system equation [k] {δ } = {F} is modified to see that the solution for {δ} comes out with the above values. This modification of system equation is called imposing the boundary conditions.

Step 7: The above 96 simultaneous equations are solved using the standard numerical procedures like Gauss elimination or Choleski's decomposition techniques to get the 96 nodal displacements.

Step 8: Now the interest of the analyst is to study the stresses at various points. In solid mechanics the relationship between the displacements and stresses are well established. The stresses at various points of interest may be found by using shape functions and the nodal displacements and then stresses calculated. The stress concentrations may be studies by comparing the values obtained at various points in the fillet zone with the values at uniform zone, far away from the fillet (which is equal to P/b_2t).

FINITE ELEMENT METHOD V/S CLASSICAL METHODS

1. In classical methods exact equations are formed and exact solutions are obtained where as in finite element analysis exact equations are formed but approximate solutions are obtained.

2. Solutions have been obtained for few standard cases by classical methods, where as solutions can be obtained for all problems by finite element analysis.

3. Whenever the following complexities are faced, classical method makes the drastic assumptions' and looks for the solutions:
(a) Irregular Shape
(b) Irregular Boundary conditions
(c) Irregular Loading

Figure-3 shows such cases in the analysis of slabs (plates).
To get the solution in these cases, rectangular shapes, same boundary condition along a side and regular equivalent loads are to be assumed. In FEM no such assumptions are made. The problem is treated as it is.

4. When material property is not isotropic, solutions for the problems become very difficult in classical method. Only few simple cases have been tried successfully by researchers. FEM can handle structures with anisotropic properties also without any difficulty.

5. If structure consists of more than one material, it is difficult to use classical method, but finite element can be used without any difficulty.

6. Problems with material and geometric non-linearities can not be handled by classical methods. There is no difficulty in FEM.

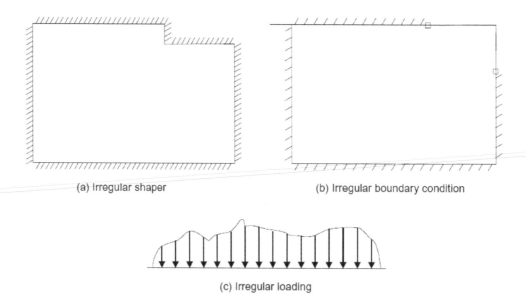

(a) Irregular shaper   (b) Irregular boundary condition

(c) Irregular loading

Figure-3. Analysis of slab

Hence, FEM is superior to the classical methods only for the problems involving a number of complexities which cannot be handled by classical methods without making drastic assumptions. For all regular problems, the solutions by classical methods are the best solutions. Infact, to check the validity of the FEM programs developed, the FEM solutions are compared with the solutions by classical methods for standard problems.

FEM VS FINITE DIFFERENCE METHOD (FDM)

1. FDM makes point wise approximation to the governing equations i.e. it ensures continuity only at the node points. Continuity along the sides of grid lines are not ensured. FEM make piecewise approximation i.e. it ensures the continuity at node points as well as along the sides of the element.

2. FDM do not give the values at any point except at node points. It do not give any approximating function to evaluate the basic values (deflections, in case of solid mechanics) using the nodal values. FEM can give the values at any point. However the values obtained at points other than nodes are by using suitable interpolation formulae.

3. FDM makes stair type approximation to sloping and curved boundaries as shown in Figure-4. FEM can consider the sloping boundaries exactly. If curved elements are used, even the curved boundaries can be handled exactly.

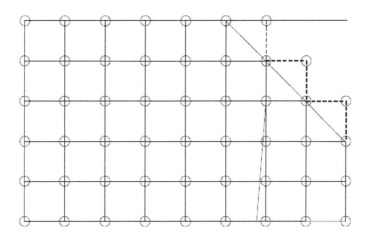

Figure-4. FDM approximation of shape

4. FDM needs larger number of nodes to get good results while FEM needs fewer nodes.

5. With FDM fairly complicated problems can be handled where as FEM can handle all complicated problems.

NEED TO STUDY FEA

Now, a number of users friendly packages are available in the market. Hence one may ask the question 'What is the need to study FEA?'.

The finite element knowledge makes a good engineer better while user without the knowledge of FEA may produce more dangerous results. To use the FEA packages properly, the user must know the following points clearly:

1. Which elements are to be used for solving the problem in hand.
2. How to discretize to get good results.
3. How to introduce boundary conditions properly.
4. How the element properties are developed and what are their limitations.
5. How the displays are developed in pre and post processor to understand their limitations.
6. To understand the difficulties involved in the development of FEA programs and hence the need for checking the commercially available packages with the results of standard cases.

Unless user has the background of FEA, he may produce worst results and may go with overconfidence. Hence it is necessary that the users of FEA package should have sound knowledge of FEA.

Warning To FEA Package Users

When hand calculations are made, the designer always gets the feel of the structure and get rough idea about the expected results. This aspect cannot be ignored by any designer, whatever be the reliability of the program, a complex problem may be simplified with drastic assumptions and FEA results obtained. Check whether expected trend of the result is obtained. Then avoid drastic assumptions and get more refined results with FEA package. User must remember that structural behavior is not dictated by the computer programs. Hence the designer should develop feel of the structure and make use of the programs to get numerical results which are close to structural behavior.

One of the main concern for using FEA applications is selection of elements, in other words discretization. The process of modeling a structure using suitable number, shape and size of the elements is called discretization. The modeling should be good enough to get the results as close to actual behavior of the structure as possible.

In a structure, we come across the following types of discontinuities:

(a) Geometric
(b) Load
(c) Boundary conditions
(d) Material.

We should try to avoid or minimize these discontinuities. If not possible then we should use split meshing at those areas.

Geometric Discontinuities

Wherever there is sudden change in shape and size of the structure there should be a node or line of nodes. Figure-5 shows some of such situations.

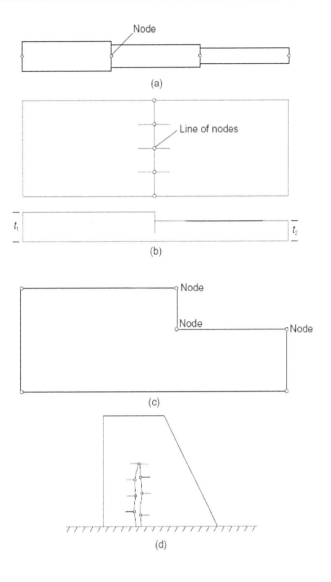

Figure-5. Geometric Discontinuity in the part

Discontinuity of Loads

Concentrated loads and sudden change in the intensity of uniformly distributed loads are the sources of discontinuity of loads. A node or a line of nodes should be there to model the structure. Some of these situations are shown in Figure-6.

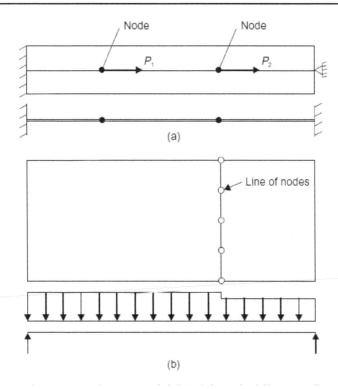

Figure-6. FEM Model and slab with different UDL

Discontinuity of Boundary conditions

If the boundary condition for a structure suddenly change, we have to discretize such that there is node or a line of nodes. This type of situations are shown in Figure-7.

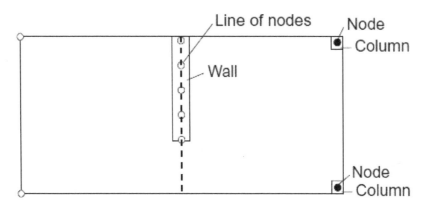

Figure-7. Slab with intermediate wall and columns

Material Discontinuity

Node or node lines should appear at the places where material discontinuity is seen; refer to Figure-8.

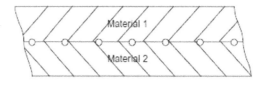

Figure-8. Material Discontinuity

REFINING MESH

To get better results the finite element mesh should be refined in the following situations:

(a) To approximate curved boundary of the structure
(b) At the places of high stress gradients.

Such a situation is shown in Figure-9.

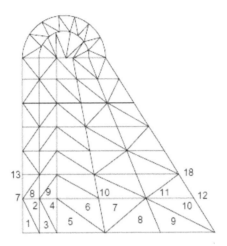

Figure-9. Refined mesh near curved boundary

Use Of Symmetry

Wherever there is symmetry in the problem, it should be made use. By doing so, lot of memory requirement is reduced or in other words we can use more elements (refined mesh) for the same capacity of computer memory. When symmetry is to be used, it is to be noted that at right angles to the line of symmetry displacement is zero.

HIGHER ORDER ELEMENTS V/S REFINED MESH

Accuracy of calculation increases if higher order elements are used. Accuracy can also be increased by using more number of elements. Limitation on use of number of elements comes from the total degrees of freedom the computer can handle. The limitation may be due to cost of computation time also. Hence to use higher order elements, we have to use less number of such elements. The question arises whether to use less number of higher order elements or more number of lower order elements for the same total degree of freedom. There are some studies in this matter keeping degree of accuracy per unit cost as the selection criteria. However the cost of computation is coming down so much that such studies are not relevant today. Accuracy alone should be selection criteria which may be carried out initially on the simplified problem and based on it element order may be selected for detailed study.

F OR S TUDENT N OTES

Chapter 16

Project on Analysis

The major topics covered in this chapter are:

- *Project*
- *Static Analysis*
- *Frequency Analysis*
- *Fatigue Test*

PROJECT

In this project,we have a real-time problem of forging die. For your understanding, I want to tell you that there are many components like gears, shafts, rings, engine covers that are made with the help of a forging machine. In a forging machine, a workpiece pre-heated at temperature of plastic deformation is placed in the bottom die and top die strikes to the workpiece. After this impact of 50kN to 1600kN on the workpiece, the workpiece takes the shape of top and bottom dies; refer to Figure-1.

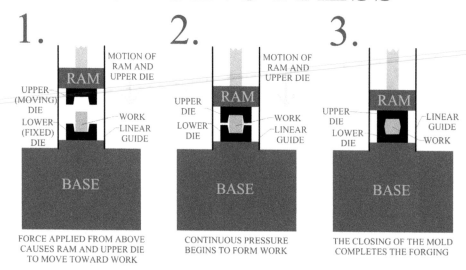

Figure-1. Forging Process

Now, we have the bottom die that need to be analyzed at **115** °C of surface temperature and force exerted on its surface is **200** kN; refer to Figure-2 and Figure-3. We need to find out the Factor of safety of the model. If it is higher then methods to reduce the cost of model. If the FoS is lower then the methods to increase it for increasing strength of the model. Also, we need to check the model for Fatigue under the cycle of 100,000 loading/unloading. After performing the analysis, we will create a more refined mesh and then we will check the results again.

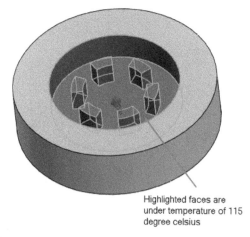

Highlighted faces are
under temperature of 115
degree celsius

Figure-2. Faces under specified temperature

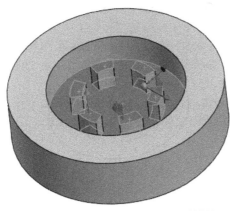

Highlighted faces are under the force of 200kN

Figure-3. Faces under the load

STARTING SIMULATION

- Open the file of die model in SolidWorks (The file is available in the Resource kit).
- Click on the **Simulation** button from the **SOLIDWORKS Add-Ins** tab in the **Ribbon**. The **Simulation** tab will be added in the **Ribbon**.
- Click on the **Simulation** tab. The tools related to simulation will be displayed.

PERFORMING THE STATIC ANALYSIS

- Click on the down arrow below **Study Advisor** in the **Simulation** tab's **Ribbon**.
- Select the **New Study** button from the list of tools displayed. The **Study PropertyManager** will be displayed.
- Click on the **Static** button from the **PropertyManager** and click on the **OK** button from the **PropertyManager**.
- Click on the **Apply Material** button from the **Ribbon**. The **Material** dialog box will be displayed as shown in Figure-4.

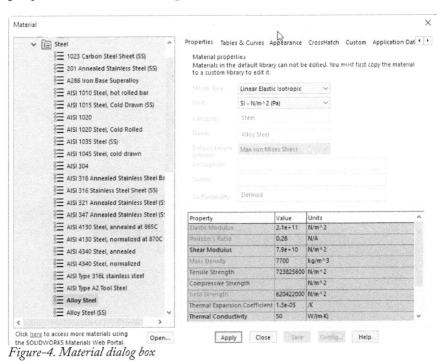

Figure-4. Material dialog box

- By default, the **Alloy Steel** material is selected from the list in the left of the dialog box. If not selected then select it.
- Click on the **Apply** button and then the **Close** button from the dialog box.
- Click on the down arrow below **Fixture Advisor** button in the **Ribbon** and select the **Fixed Geometry** button from the list.
- The **Fixture PropertyManager** will be displayed.
- Select the bottom face of the model to fix it; refer to Figure-5.

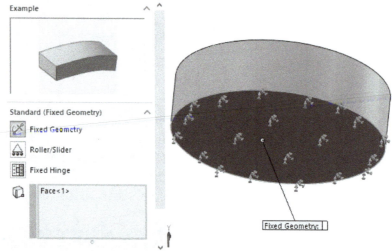

Figure-5. Face to be fixed

- Click on the **OK** button from the **PropertyManager**.
- Click on the down arrow below **External Loads Advisor** button in the **Ribbon**. The list of tools will be displayed.
- Click on the **Temperature** button from the list. The **Temperature PropertyManager** will be displayed.
- Click on the **Select all exposed faces** button from the **PropertyManager**. All the faces will get selected.
- Press and hold the **CTRL** key and select the bottom, side and top face of the die; refer to Figure-6. The faces will get deselected.

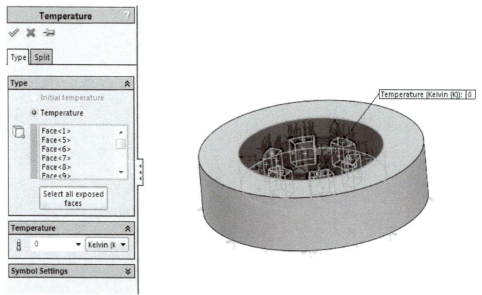

Figure-6. Faces selected for temperature

- Click on the **Unit** drop-down in the **Temperature** rollout and select the **Celsius (°C)** option.
- Click in the **Temperature** edit box and specify the value as **115**.
- Click on the **OK** button from the **PropertyManager**.
- Again, click on the down arrow below **External Loads Advisor** button in the **Ribbon**.
- Select the **Force** button from the list of tools. The **Force PropertyManager** will be displayed.
- Select the faces visible from the top of the model; refer to Figure-7.

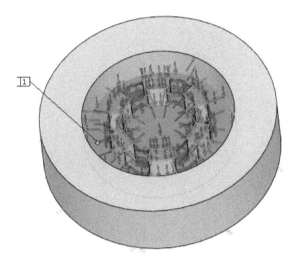

Figure-7. Faces for applying force

- Click in the **Force Value** edit box and specify the value as **200000**N.
- Select the **Total** radio button from the **PropertyManager** and click on the **OK** button.
- Click on the **Run This Study** button from the **Ribbon**. After the analysis is solved the result will be displayed; refer to Figure-8.

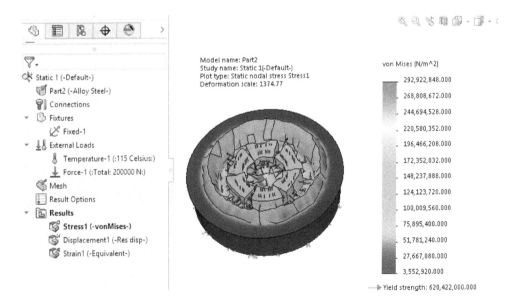

Figure-8. Result of static analysis

- Right-click on the **Results** node in **Analysis Manager** and select the **Define Factor Of Safety Plot** option from the shortcut menu displayed.
- The **Factor of Safety PropertyManager** will be displayed; refer to Figure-9.

Figure-9. Factor of Safety PropertyManager

- Click on the **OK** button from the **PropertyManager**. The **Factor of Safety** plot will be displayed; refer to Figure-10.

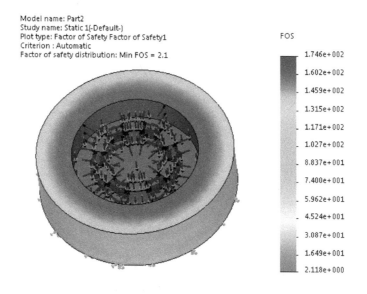

Figure-10. Factor of Safety plot

PERFORMING THE FREQUENCY ANALYSIS

- Click on the down arrow below **Study Advisor** and select the **New Study** button from the list displayed.

- Click on the **Frequency** button from the **Study PropertyManager** and click on the **OK** button from the **PropertyManager**. The tools related to frequency analysis will be displayed.
- Click on the **Apply Material** button from the **Ribbon**. The Material dialog box will display.
- Select the **Alloy Steel** from the left of the dialog box.
- Click on the **Apply** button and then on the **Close** button from the dialog box.
- Fix the bottom face of the die as done in the previous analysis by using the **Fixed Geometry** button.
- Click on the **Run This Study** button from the **Ribbon** to perform the analysis.
- After the analysis is complete, the natural frequencies of the model will be displayed; refer to Figure-11.

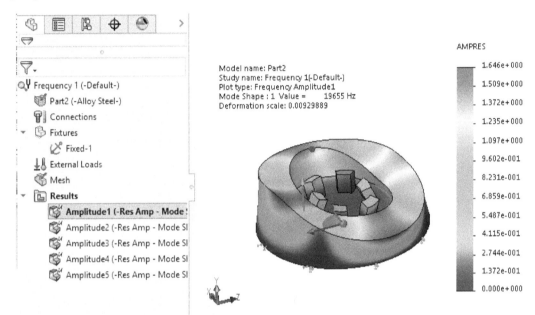

Figure-11. Result of frequency analysis

- Make sure that the frequency of the machine vibration is not matching with the natural frequencies of the model.

PERFORMING FATIGUE STUDY

- Click on the down arrow below **Study Advisor** in the **Ribbon**. List of the tools will be displayed.
- Click on the **New Study** tool. The **Study PropertyManager** will be displayed.
- Click on the **Fatigue** button from the **PropertyManager** and click on the **OK** button from the **PropertyManager**. The tools related to **Fatigue analysis** will be displayed.
- Right-click on **Loading (-Constant Amplitude-)** in the **Analysis Manager** and select the **Add Event** option; refer to Figure-12. The **Add Event (Constant) PropertyManager** will be displayed; refer to Figure-13.

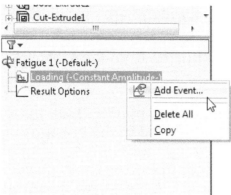

Figure-12. Add Event option

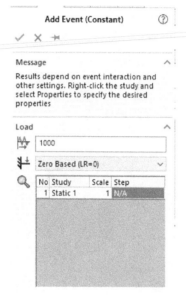

Figure-13. Add Event Property-Manager

- Click in the **Cycles** edit box and specify the value as **1000**.
- Click on the **Loading Type** drop-down and select the **Zero Based (LR=0)** option from the drop-down list.
- Click on the **OK** button from the **PropertyManager** to add the event. Note that name of part will be added as a new node in the **Analysis Manager**.
- Right-click on the name of the part and select the **Apply/Edit Fatigue Data** option; refer to Figure-14. The **Material** dialog box will be displayed with the **Fatigue SN Curves** tab activated; refer to Figure-15.

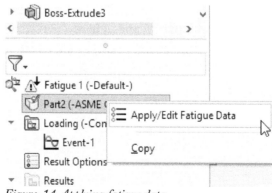

Figure-14. Applying fatigue data

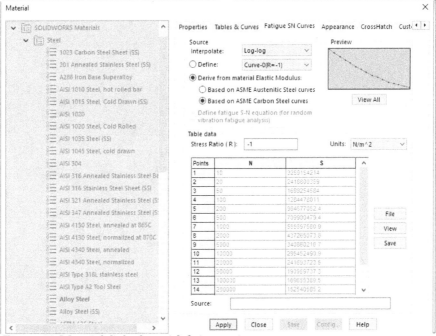

Figure-15. Material dialog box with fatigue options

- Click on the **Derive from material Elastic Modulus** radio button and then click on the **Based in ASME Carbon Steel curves** radio button. (Due to selection of these radio buttons, the SN curve of a carbon steel will be used which is derived from its elasticity curve.)
- Click on the **Apply** button and then click on the **Close** button from the dialog box. Now, we are ready to perform the analysis.
- Click on the **Run This Study** button from the **Ribbon**. The analyzing will start.
- The results of the analysis will be added in the **Results** node.
- Expand the node and double-click on the results to display the plot; refer to Figure-16.

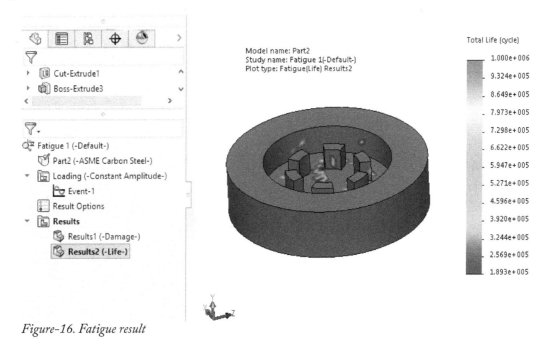

Figure-16. Fatigue result

- Right-click on the **Results** node and select the **Define Fatigue Plot** option; refer to Figure-17. The **Fatigue Plot PropertyManager** will be displayed; refer to Figure-18.

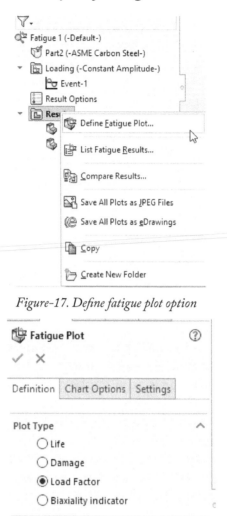

Figure-17. Define fatigue plot option

Figure-18. Fatigue Plot PropertyManager

- Click on the **Load Factor** radio button and click on the **OK** button from the **PropertyManager**. The load factor plot (in other words, **Factor of Safety** for fatigue) will be displayed; refer to Figure-19.

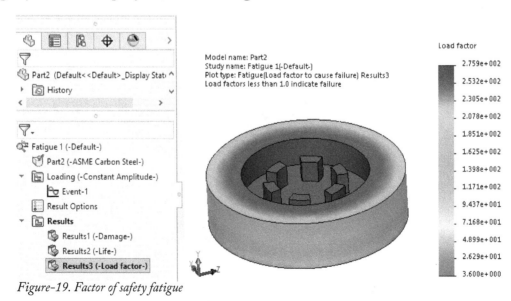

Figure-19. Factor of safety fatigue

- From the scale, the you can find that the minimum factor of safety of the fatigue study is **3.6**.

Since our **Factor of Safety** for the problem is optimum, so we do not required to perform the **Design Study** as performed in previous chapters. Although, you can perform the Design Study to find a better product.

Now, I have a little work for you:
Go back to **Static Study** analysis by clicking on the **Static 1** tab at the bottom bar of the modeling area. Note that the **Factor of Safety** is **2.1** for this analysis. Now, Right-click on the **Mesh** node in the **Analysis Manager**; refer to Figure-20.

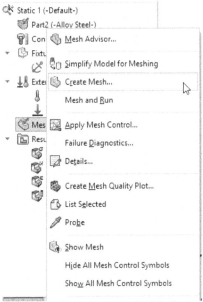

Figure-20. Shortcut menu for Mesh

Select the **Create Mesh** button, make the meshing extremely fine and then re-run the static analysis. Now, check the **Factor of Safety** of the model. Is it the same or different. If different then why? Ask your trainer.

FOR STUDENT NOTES

FOR STUDENT NOTES

Practice Questions

Topics Covered

The major topics covered in this chapter are:

- *Practice Questions*

PROBLEM 1

Sometimes we see people fighting and in some extreme cases, fighting with the help of Baseball bats. We are not here to discuss concepts of fighting but once the fighting is over, everybody is busy to see effect of baseball bat on humans. Due to unknown elastic property of human body, we can not perform analysis to find effect of baseball bat on humans, but here we can check the effect of force exerted on the baseball bat due to some freaks; refer to Figure-1.

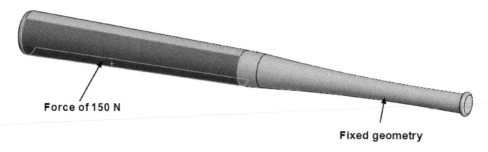

Force of 150 N

Fixed geometry

Material: Aluminium alloy 1060

Figure-1. Problem 1

We assume it to be a static analysis. Check whether the bat will break or not.

PROBLEM 2

The wall of a house is 7 m wide, 6 m high and 0.3 m thick made up of brick. The thermal conductivity of brick is k = 0.6 W/m.K. If the inside temperature of wall is 16 °C and outside temperature is 6 °C then find out the heat lost from inside wall to outside wall in a steady state condition. The model is given in Figure-2.

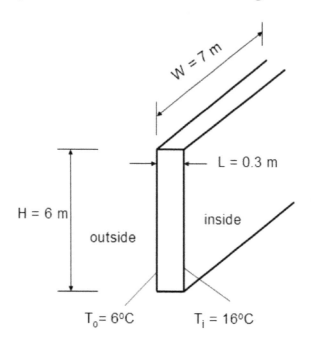

W = 7 m

L = 0.3 m

H = 6 m

inside

outside

$T_0 = 6°C$ $T_i = 16°C$

Figure-2. Problem 2

Note that to perform this thermal analysis, you need to create a new material with thermal properties as follow:

Thermal Conductivity = 0.6 W/(m.K)
Specific heat = 1214 J/(kg.K)
Mass density = 2405 kg/m^3

PROBLEM 3

An axial pull of **35kN** is applied at the ends of a shaft as shown in Figure-3. If the material is **Alloy Steel** then find the total elongation in the shaft.

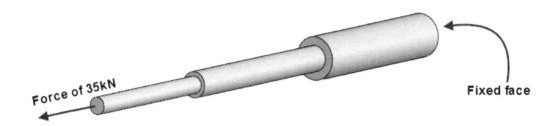

Figure-3. Problem 3 shaft

PROBLEM 4

A cast iron link is required to transmit a steady tensile load of 45kN. The initial boundary conditions are given in Figure-4.

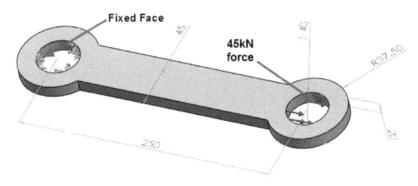

Figure-4. Problem of link

Find out the optimum thickness of link so that factor of safety is 3.

PROBLEM 5

A part of car jack assembly is displayed in Figure-5 with its boundary conditions. Optimize the assembly with respect to material cost. Note that all the contacts are no penetration as you find in a real car jack.

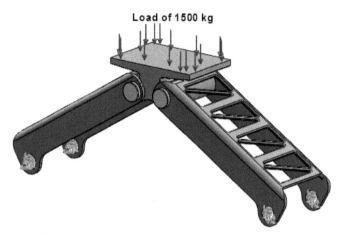

Figure-5. Problem 5

PROBLEM 6

An assembly of cooling fins and CPU is displayed in Figure-6 with boundary conditions. The bulk temperature is 300K and operating temperature band for CPU is 304K to 354 K. Check if fins are enough to dissipate the heat. If not then perform modifications in fin to dissipate heat. Note that fins and heat sink are perfectly bonded for heat transfer.

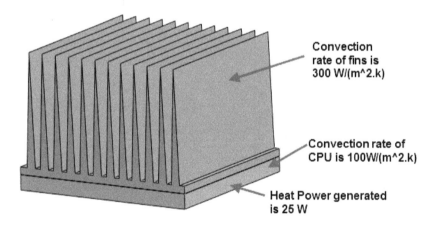

Figure-6. Problem 6

Index

OTHER BOOKS BY CADCAMCAE WORKS

Autodesk Inventor 2018 Black Book

Autodesk CFD 2018 Black Book

Autodesk Fusion 360 Black Book

Basics of Autodesk Inventor Nastran 2020

AutoCAD Electrical 2020 Black Book
AutoCAD Electrical 2018 Black Book
AutoCAD Electrical 2017 Black Book

SolidWorks 2020 Black Book
SolidWorks 2019 Black Book
SolidWorks 2018 Black Book
SolidWorks 2017 Black Book
SolidWorks 2016 Black Book

SolidWorks Simulation 2020 Black Book
SolidWorks Simulation 2019 Black Book
SolidWorks Simulation 2018 Black Book
SolidWorks Simulation 2017 Black Book

SolidWorks Flow Simulation 2020 Black Book
SolidWorks Flow Simulation 2019 Black Book
SolidWorks Flow Simulation 2018 Black Book

SolidWorks Electrical 2020 Black Book
SolidWorks Electrical 2019 Black Book
SolidWorks Electrical 2018 Black Book
SolidWorks Electrical 2017 Black Book

Mastercam X7 for SolidWorks 2014 Black Book
Mastercam 2017 for SolidWorks Black Book

Creo Parametric 6.0 Black Book
Creo Parametric 5.0 Black Book
Creo Parametric 4.0 Black Book

ETABS 2016 Black Book

Autodesk Revit 2020 Black Book